FAST REACTOR SAFETY

JOHN GRAHAM

Westinghouse Advanced Reactors Division
Madison, Pennsylvania

A C A D E M I C   P R E S S   New York and London   1971

ACADEMIC PRESS, INC.
111 Fifth Avenue, New York, New York 10003

United Kingdom Edition published by
ACADEMIC PRESS, INC. (LONDON) LTD.
24/28 Oval Road, London NW1 7DD

LIBRARY OF CONGRESS CATALOG CARD NUMBER: 71-154370

PRINTED IN THE UNITED STATES OF AMERICA

FAST REACTOR SAFETY

NUCLEAR SCIENCE AND TECHNOLOGY
A Series of Monographs and Textbooks

CONSULTING EDITOR

V. L. PARSEGIAN

Chair of Rensselaer Professor
Rensselaer Polytechnic Institute
Troy, New York

1. John F. Flagg (Ed.)
   CHEMICAL PROCESSING OF REACTOR FUELS, 1961

2. M. L. Yeater (Ed.)
   NEUTRON PHYSICS, 1962

3. Melville Clark, Jr., and Kent F. Hansen
   NUMERICAL METHODS OF REACTOR ANALYSIS, 1964

4. James W. Haffner
   RADIATION AND SHIELDING IN SPACE, 1967

5. Weston M. Stacey, Jr.
   SPACE-TIME NUCLEAR REACTOR KINETICS, 1969

6. Ronald R. Mohler and C. N. Shen
   OPTIMAL CONTROL OF NUCLEAR REACTORS, 1970

7. Ziya Akcasu, Gerald S. Lellouche, and Louis M. Shotkin
   MATHEMATICAL METHODS IN NUCLEAR REACTOR DYNAMICS, 1971

8. John Graham
   FAST REACTOR SAFETY, 1971

In preparation:

   Akinao Shimizu and Katsutada Aoki
   APPLICATION OF INVARIANT EMBEDDING TO REACTOR PHYSICS

To Claire

CONTENTS

FOREWORD

This book is a step along the arduous path of developing the technology of nuclear safety as a fully recognized discipline with quantitatively defined standards and an accepted methodology of analysis, design, and test interpretation. Completion of this task is needed to stabilize the nuclear power industry and eliminate the elements of unfounded opinion, and even emotion, which too often characterize public discussions of nuclear power plant safety, particularly in the application of fast power reactors.

A book devoted to nuclear safety as a special and separate skill raises immediately a conflict for engineers who have been associated with the design, construction, testing, and operation of a total nuclear power plant, a conflict caused by the conviction that if the designer does not provide reliability and safety in each of the components and systems for which he is responsible, that plant will not be reliable or safe. In fact, the existence of a separate skill in nuclear safety opens up a potential risk that the system or component designer will leave the matter of safety to this new specialist and in so doing, miss providing the safety features which only he can successfully assure. Guidance in resolving this conflict can be obtained by examining other fields in which I feel a similar dilemma has been faced and resolved. Let us look at an analogous case.

The development of high performance equipment in fields such as aviation, space, modern structures, and of course, nuclear power itself, has required the development of specialized skills in stress analysis and mechanics. This technical specialty is so demanding that the full careers of highly trained people must be exclusively devoted to it. Yet, their skills must be utilized by the designer and highly effective technical communication between the designer and the sophisticated practitioner of mechanics is necessary. The success of achieving such communication has meant the success or failure of many projects. Similarly, the skill and knowledge of the highly trained metallurgist, pursuing his experiments and analysis as a separate endeavor, must be absorbed by the designer to ensure the proper use of materials in the component he is designing. Here again the skill must be developed independently because of its demanding nature, but the designer must utilize results of such development if he is to make a success of his project.

Nuclear safety, although not as well recognized today in the curricula of universities as is the study of mechanics or metallurgy, fulfills an analogous role. This skill must be recognized in a manner similar to that of mechanics and metallurgy. Further, it should not simply be restricted to nuclear safety but the basic methodology must also be applied to safety of all modern, massive, "high-technology" devices. The forerunner of this book, in fact, has been a set of lecture notes developed by the author for a course given as part of the nuclear engineering curriculum at Carnegie–Mellon University.

The transfer of nuclear safety skills to the designer must involve more than discussion, consultation, and teaching. It requires the definition and acceptance of safety design criteria, and the development of design methodology, to which a substantial portion of this book is devoted. In modern application of analysis and design techniques, the computer code is ever present as a specific device for this methodology transfer. It should be apparent from the discussions in the text, however, that many of these codes are still in a state of development and should not be used by the designer except with a full appreciation of their limitations. A third, and ultimately most important, means of transfer of safety skills is the incorporation into design of the actual experience gained by safety experiments and by plant operation. The former must be incorporated through the comparison of design methods against definitive experiments and the subsequent improvement of these methods to more accurately portray the accident, or the accident initiators. This area of accident analysis verification is still quite weak, as indicated by the relative paucity of such experimental comparisons the author of this book is able to make. It is not clear that there is enough emphasis today, in the planning of experiments and in the development of analytical methods to interpret such experiments, on satisfying the need of the designer to have design methods which can be compared with such experiments and then applied with prudent extrapolation to his reactor design. In the latter case of the incorporation of plant operational experience, the methods of reliability analysis reviewed in the book are a main vehicle of transfer. Fortunately, there are many common features between fast reactors and the present generation of thermal reactors, particularly in the areas of reactor protection and accident prevention. As a result, the operational data accumulated on present day nuclear plants and their components can be incorporated into the reliability analyses of the fast reactor designer.

Attention by specialists in nuclear safety to the above means of transfer of their skill will prevent the "separation" of design from safety expertise and will, in addition, assure the proper balance in all the activities needed to design, construct, and operate a safe nuclear power plant.

One must be concerned about the tendency of pursuing safety technology for its own sake rather than in close coordination with design, so as to overemphasize the study of highly unlikely accidents and means of coping with their consequences. Such overemphasis inevitably results in a less than desired effort on (a) accident prevention through the study of accident initiators and design approaches to eliminate them, (b) the study of relatively probable accidents and the design of protective systems to assure that no significant loss in reactor or plant integrity will occur as a result of the accident, and (c) the definition of a highly reliable design of the plant systems and components which achieves safety through such reliability.

The importance of keeping the proper balance is underlined by the realization that I believe is held strongly by those experienced in nuclear operation: namely, that a functionally complex or awkward plant or one with unreliable components is an accident-prone plant. No matter how many engineered safeguards systems there are on such a plant to cope with accidents, an accident-prone plant is not a safe plant. Further, the goal of designing a reliable plant, if rigorous standards, quality assurance procedures, and thorough preoperational testing and diagnostic in-service testing are employed, is more assured than the counterpart goal of designing reliable engineered safeguards. Experience tells us that when we need a system to provide protection against accidents we should, wherever possible, utilize equipment which is a continued operational part of the plant and thus will receive the ultimate test of reliability—continuous performance.

One other important contribution this book can make, particularly for those of us devoting our energies to the development of the fast reactor, is to bring fast reactor safety into context. Those engaged in fast reactor development do not find substance in the popular image that there is something especially dangerous about a fast reactor. I can do no better in this respect than to quote from a statement by F. R. Farmer in his Foreword to a summary report[†] on fast reactor safety published last year with which I heartily agree. He states,

> There has been a prolonged and excessive preoccupation with the apparent differences between fast and thermal systems and, for many years, a particular interest in the explosive disruption of a fast reactor core.... In general, it is shown that differences exist between all reactors and the fast reactor is not notable in this respect, nor in respect to explosive potential.... The facts which may initiate the various modes of destructive failure will be different between fast and thermal systems and will call for different methods of detection, but it is not obvious that one deserves greater effort or leads to greater concern than the other.

This book, by examining as quantitatively as our present technological capability permits the specific features of safety of the fast system, makes it clear that the fast reactor can be designed to be safe. The sodium coolant in the Liquid Metal Fast Breeder Reactor, for example, has significant advantages from a safety standpoint. The system can operate essentially at atmospheric pressure where the maximum pressure is a result only of the pump head. The boiling point of sodium is 500–600°F above the peak operating temperature range of the sodium coolant, largely eliminating concern about boiling in the reactor core. The historic concern about the short, prompt neutron lifetime of a fast reactor has been allayed for the most part by the finding that the rate of power increase caused by a reactivity addition in excess of prompt critical is limited to safe levels by Doppler feedback. The historic concern relative to a core disruptive accident is being brought into perspective by focusing attention both in analysis and experimentation on defining the quantitative features of this accident rather than the qualitative upper limit possibilities of it. Effort is also being placed on study of the initiators of the core disruptive accident so that design steps can be taken to eliminate such initiators. Of particular importance in this regard is the experimental program on fuel element failure propagation which must either establish that failure propagation is limited to safe levels

---

[†] F. R. Farmer *et al.*, An Appreciation of Fast Reactor Safety (1970), AHSB(S)-R-188. Authority Health and Safety Branch, United Kingdom Atomic Energy Authority, Risley, Warrington, Lancashire, England.

or show the way to fuel and fuel assembly design modifications to limit propagation to safe levels. Thus, there is no unique barrier to providing this power generating device that opens up vast natural fuel resources to mankind without exposing him to undue risks to his life and property.

It would be consoling to be able to say that this book completes the task of placing fast reactor safety beyond opinion and emotional issues, and reduces the entire subject to that of professionally recognized skills and standards. It does not because there are still substantial technical developments in nuclear safety which have yet to be completed, both in the analytical and experimental areas. Lest we be discouraged, however, we must remember that this is the first major industrial technological enterprise which is being subjected to rigorous, professional development of its safety characteristics.

A final word on one other aspect of the book's contribution. There has been an increasing call in fast reactor breeder development to establish safety criteria and licensing standards on an international basis. Such a step would better assure the coordinated and constructive response of all workers in the field to achieve the best set of such criteria and standards. In addition, artificial barriers involving licensing requirements set up along national lines would hopefully be reduced or even eliminated. This book does not directly address this issue. But is assists the process of achieving a more uniform international approach by presenting the licensing position as it stands today in the U.S. Similar presentations of licensing positions in Britain and Germany are also making a contribution to this international cooperation. John Graham, the author of this book, has carried out in his own career a significant amount of safety work across national boundaries, having participated heavily in the fast reactor safety program in Great Britain and now, more recently, in the safety analysis and licensing activities associated with the liquid metal fast breeder reactor program in the United States.

> J. J. TAYLOR
> *General Manager*
> *Breeder Reactor Divisions*
> *Westinghouse Electric Corporation*

PREFACE

In the design of a reactor and its associated plant systems the involvement of safety engineering is total. Safety appears in the selection of design concepts, the design itself, and as an evaluation of the design. The safety of the plant being the subject of the Preliminary and Final Safety Analysis Reports which are submitted in support of plant construction and operating licenses, in the final analysis may be an overriding consideration in the production of the nuclear power system.

Because of the total involvement of safety engineering at all stages of the design and in all sections of the design, the field of reactor safety requires an immense range of technical skills. In the past, reactor safety engineers have been drawn from the ranks of those who are specialists in one single area of reactor technology and who have attained experience enough to recognize the required wide background necessary for the assessment of safety. Nevertheless in any safety group it has always been necessary to cover the whole range of skills required by providing for a number of personnel with different backgrounds. Ideally these engineers should be subjected to an overall reactor safety course to enable them to think and speak in a consistent manner with respect to safety.

Modern university curricula instruct in sccialist skills each of which may be limited to a single technical field. It is therefore important to superimpose upon the basic university instruction, preferably as early in the educational process as possible, applied safety courses which emphasize and show the interactions among a range of skills drawn from many different fields. The short extra-curricular course is not likely to provide the balance and depth of understanding needed for valid training in safety.

The present volume is intended to fill the need for a university text for reactor safety applied to fast reactors in general and applied to liquid-metal-cooled fast breeders in particular. Liquid-metal-cooled fast breeders are the favored fast reactor concepts for the major nuclear countries of the world and therefore the emphasis is pertinent to our future needs.

One may ask why such a volume is not first devoted to the present generation of nuclear power plants—the thermal pressurized and boiling light water systems. The answer is that undoubtedly such a volume is needed but that the author's experience leads more directly to the fast reactor. However it is worth noting that all but Chapters 4 and 5 of the present volume also apply to the thermal reactor system and a thermal reactor safety engineer will also find the book of use.

The book is intended as a university text for graduates and undergraduates in nuclear engineering who are attending courses in reactor safety. Safety engineering encompasses mathematics, thermal hydraulics, fluid dynamics, control theory, logic analysis, nuclear physics, structural mechanics, stress analysis, metallurgy, licensing regulations, meteorology, health physics, and a host of other technical fields. It is therefore necessary to require that the student and the reader should possess certain prerequisites. The minimum should be a basic knowledge of differential calculus, nuclear reactor theory, and some heat transfer and fluid dynamics.

The text of the book has been used in teaching the subject of fast reactor safety at Carnegie–Mellon University, Pittsburgh, and the feedback from the presentation there has greatly improved the book. I am grateful for the comments received from my students.

The text also owes much to work performed by other organizations including: the International Atomic Energy Agency, the United Kingdom Atomic Energy Authority, the Atomic Energy Commission, Argonne National Laboratory, Oak Ridge National Laboratory, the British Nuclear Energy Society, the Institution of Mechanical Engineers, the Liquid Metal Engineering Center, the American Society of Mechanical Engineers, the American Society for Testing Materials, the Boeing Company, and Westinghouse Electric Corporation. Many individuals have also helped the work by their criticisms as well as by material contributions, and John Zoubek should be singled out for his assistance in several sections of the book.

I am particularly grateful to Dr. R. G. Cockrell for his continued help and constructive advice and to my wife for her encouragement, enthusiasm, and tolerance of the curious habits of someone hampered by the lack of sufficient hours in the day.

LIST OF SYMBOLS

Neutron Kinetics

n Number of neutrons of energy E and position $\mathbf{r}$ at time t; $n(\mathbf{r}, E, t)$

ϕ Flux (nv) where the average neutron velocity is $v(E)$

k_∞ Multiplication factor $(k_\infty = p\varepsilon\eta f)$

k_{eff} Effective multiplication factor

Σ Nuclear macroscopic cross section

$\nabla^2\phi$ Flux curvature

D Diffusion coefficient

δk Excess multiplication $(\delta k = k_{\text{eff}} - 1)$

C_i Concentration of delayed neutron precursors of group i

β Delayed neutron fraction

β_i Delayed neutron fraction of group i

λ_i Delayed neutron time constant for group i

f Thermal utilization

l^* Mean neutron lifetime

p Resonance escape factor

ε Fast fission factor

η $\nu\Sigma_{\text{fission}}/\Sigma_{\text{a}}$

ν Number of neutrons produced per fission

L Diffusion length $(L^2 = D/\Sigma_{\text{a}})$

B Buckling

α_i Reactivity feedback coefficient for temperature feedback i

ϱ Reactivity $(\varrho = \delta k/k)$

ω Reactor period

t	Time	μ	Viscosity
τ	Fermi age	f	Void fraction
		n	Node in a spatial finite difference model

Thermal Hydraulics

k	Thermal conductivity	η	Bulk boiling heat transfer parameter
t	Time		
T	Temperature	G	Gravitational terms in boiling equations
H	Heat source per unit volume		
		τ	Time constants and, in Eq. (1.50), the friction factor
c	Specific heat		
z	Linear distance		
r	Radial distance	BB	Boiling boundary
R	Fuel pin radius	γ	Ratio of specific heats
A	Cross-sectional area	V	Volume, especially void volume
v	Velocity		
m_i	Mass ($m = \varrho A$) of material i	q	Heat flux
ϱ_i	Density of material i		
h	Heat transfer coefficient [or enthalpies in Eqs. (1.48) *et seq.*]	**Miscellaneous**	
d_e	Equivalent diameter	I	Iodine concentration
p	Channel perimeter	X	Xenon concentration
R_e	Reynold's number ($R_e = d_e \varrho v/\mu$)	σ_X	Xenon production microscopic cross section
P_r	Prandtl number ($P_r = c_p \mu/k$)	B	Burn-up
		$N(r)$	Spatial distribution of power
N_u	Nusselt number ($N_u = h d_e/kA$)	$D(r)$	Worth function
		θ	Angular coverage
		ε	Mechanical strain

Other symbols used infrequently are defined when first used in text.

Subscripts

		F	Fast
		E	Epithermal
a	Total absorption cross section	T	Thermal
fuel	Fuel absorption cross section	B	Blanket
		ap	Parasitic absorption
fission	Fission cross section	sat	Saturated value
		0	Initial value

f	Fuel	u	Ultimate stress or strain
surf	Surface temperature	y	Yield stress or strain
c	Coolant		
cd	Cladding		

Superscripts

av	Average		
s	Structure	T^*	Absolute temperature
in	Inlet values	$\bar{T}$	Average temperature
X	Xenon	Q^*	Threshold energy
I	Iodine	h'	Modified heat transfer
l	Liquid sodium		coefficient
ν	Sodium vapor	l^*	Mean neutron lifetime to
i	General suffix for material i		differentiate from the effective neutron lifetime l

CHAPTER 1

SAFETY EVALUATION METHODS

1.1 Purpose and Plan for Safety

We have to place safety in perspective by first outlining the position of nuclear power in the world power production program and then showing how fast reactors in particular might enter into consideration. Safety is not an absolute science, and it can only be viewed in relation to the questions of how much safety is needed and how much one is prepared to pay for it. These questions will have to be answered ultimately by society itself. For the present we will outline the choices and present a personal answer.

1.1.1 ENERGY DEMAND

Before 1800 (Table 1.1), all of the world's power needs were satisfied by wood, wind, and water, all of which were energy sources that could be renewed.

Simultaneously with the growth of industry, indeed as an initiator for much of the growth, energy from coal deposits began to supply an ever-increasing proportion of the world's requirements at the end of the 19th

TABLE 1.1

WORLD SOURCES OF ENERGY: 1800–2000 [a]

World population (billion)	Energy demand (10^5 MW)	Period	Wood, Wind, Water,% [b]	Coal,% [c]	Gas,% [c]	Oil,% [c]	Nuclear
		Before 1800	100				
		1830		Start			
		1870	75	25			
1.5	4.3	1900	10	90			
2.1	9.4	1930			Start		
3.8	21.3	1970	4	20	31	44	<1
6.3	50.0	2000	3	16	32	32	18[d]

[a] See reference (2).
[b] Renewable sources of energy.
[c] Exhaustible sources of energy.
[d] In the USA the figure is predicted to be over 25%.

century. This growth in coal use continued until it supplied 90% of the world's requirements by the beginning of this century.

However, the growth of the population in recent times has a doubling time of about 40 years, and the energy demand rises even faster with a doubling time of about 25 years. Thus the supply of energy by coal could not keep pace with the demand.

A partial solution was found around 1930 when the problem of the transportation of naturally occurring oil and gas was solved, so that these two new sources of energy began to supply an ever increasing proportion of the amount required. In particular, the growth of modern transport absorbed a large proportion of the energy produced from oil. Thus by 1970 coal was supplying about a fifth of the demand with coal and oil each supplying in the neighborhood of one-third to two-fifths. The amount of energy supplied by wood, wind, and water had undoubtedly increased with the advent of new hydroelectric plants but nevertheless it supplied a very small fraction of the total (Fig. 1.1).

Unfortunately coal, gas, and oil are all exhaustible sources of energy. With the power demand rising on an increasing scale it behooves us to look to the future. The United States has the highest energy demand in

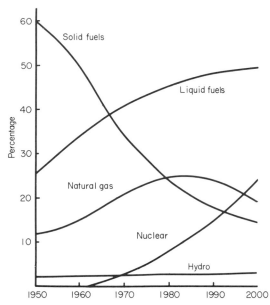

Fig. 1.1. Projected changes in the percentage shares of primary fuels in meeting the world's energy requirements (*1*).

the world, but the growth of that demand is relatively small compared to that of other nations. Figure 1.2 shows that the growth rate is larger for countries with smaller gross national products, as one might expect. This means that as the world becomes more and more affluent, the power requirements will rise to extremely high values.

Nevertheless our use of coal, oil and gas has been so rapid already that coal is likely to be exhausted in the world by about the year 2200 (*2*) with gas and oil not far behind. Thus the energy demand will not be satiable.

Certain solutions open to us are: the curtailment of population growth; the curtailment of energy demand per head of population; and a new source of energy which is renewable.

As I have said the final answer will be given by society, but I would suggest that the first two courses of action will only partially answer the problem. A new energy source will have to be made available.

Solar energy, energy derived from the fusion of light elements, tidal energy, and nuclear energy derived from fission are all candidates. Indeed Table 1.1 shows that nuclear energy is already supplying a significant proportion of the demand.

Research in the conversion of solar and fusion energy has not yet shown that these will ever be adequate sources of power for an industrial com-

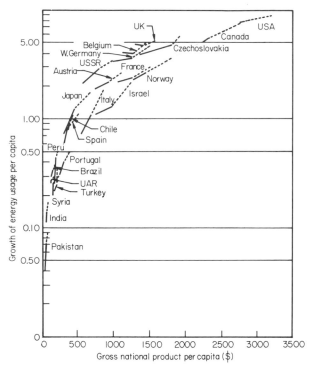

Fig. 1.2. Per capita growth of energy usage as a function of the gross national product per capita (*1*). Legend: — 1955–1960; – – – 1960–1965.

munity, while tidal power is limited in locality and availability. Nuclear power alone seems to provide the answer and by the year 2000 it is expected to provide nearly a fifth of the total needs.

Moreover, fast reactors are essentially a renewable form of energy. So while thermal reactor systems are undoubtedly short term solutions, it is only the fast breeder which will supply the long term needs in an inexhaustible form. We are thus led to the conclusion that fast reactor power plants *will* necessarily be built.

1.1.2 THE ROLE OF SAFETY

Power reactor safety has several objectives: protection of the plant against damage; protection of the public; and the presentation of evidence of safety. These objectives must be viewed against a background in which a new source of energy that is vitally needed to solve the world's power problems was born during wartime. The political use to which nuclear power was put has done a considerable amount of damage to its civilian use. Thus the final safety objective is a vital one.

In the long run, the presentation of safety is concerned with the education of society to nuclear power in general and fast reactors in particular, while in the short run it is concerned with the presentation of the safety analysis of a particular power plant.

The protection of the public is concerned with obtaining an assurance that there can be no plant or system disturbance which could ever result in the release of a significant quantity of fission products from the plant site. Such a subject also involves siting policies as part of the safety considerations.

However, the main objective is the design objective in which the plant (and of course the public) is protected against damage. The design objectives for any power plant will include safety, economical operation, reliability as well as flexibility, ease of operation, and compatibility with other power sources; but safety is a primary one.

1.1.3 THE TEXT

This book therefore presents safety as an embryo science by outlining the analytical tools which are available to the safety engineer, by presenting the experience which has been obtained using fast reactors to date, and finally by posing the questions which remain unanswered. This is done within the framework of designing a fast reactor power plant and of taking it through the regulatory process.

Chapter 1 presents the methods of safety evaluation, Chapter 2 outlines possible disturbances to the system and their interactions. Chapters 3–6 take the reader through the design and licensing process, from the establishment of safety criteria, through special fast reactor considerations in safety, the barrier concept of safety, and the final presentation of the plant safety to the regulatory bodies and the public. For a detailed discussion of fast reactor technology, overall reactor safety, and a review of current safety practices, the reader is recommended to use the general references listed at the end of this chapter to supplement the present text.

1.2 Neutron Kinetics

It is necessary to perform kinetic calculations on any reactor system throughout the design of that system from its initial conception through to its commissioning. In the initial assessment period calculations may necessarily be rather crude because of the uncertainties in most of the parameters. Average models may be used to identify the particular operating and safety philosophy to be followed, to define further studies to be performed, and to define failure mechanisms and possible ranges of operation. Many of these studies will have design implications and some will have safety repercussions.

Once the design of the system starts, it is possible to carry out more elaborate calculations. These calculations will investigate transient conditions of the system, both in normal operation and in accident disturbances. Transient conditions will include thermal distributions, dose levels, and the value of safety margins to be provided. The stability of the system will be assessed. It may be necessary to have a number of purpose-built mathematical models to investigate all these effects. During this stage kinetics calculations will help to define control procedures and safety system characteristics. As the work proceeds and the design hardens, the model parameters are updated continuously, culminating in a set of final runs to provide the safety documentation of the system.

Thus the kineticist works as a member of the assessment and design teams together with steady-state physicists, design engineers, and safety engineers.

1.2.1 NEUTRON MULTIPLICATION[†]

The heart of the kinetics model is the representation of the neutron behavior. This behavior is very similar, whether the neutrons concerned are at thermal energy or whether they are, on the average, fast (i.e., more energetic). Most university courses expand nuclear reactor theory from the point of view of thermal reactor systems and therefore the following text, for the convenience of its readers, does the same. The points at which the fast reactor differs from the thermal system are noted and this emphasizes where and how fast reactor systems differ from their thermal counterparts.

1.2.1.1 *Multiplication Factor*

The behavior of average neutrons is governed by certain probability functions called cross sections. These define the probabilities of absorption, scattering, and of fission within the given system.

Assuming an initial flux of thermal neutrons ϕ in a thermal reactor core we can calculate the multiplication of neutrons as follows:

[†] See Glastone and Edlund (3).

Thermal flux ϕ

Number absorbed into the fuel $\Sigma_{\text{fuel}}\phi/\Sigma_a = f\phi$

Number producing fission $f\phi\,\Sigma_{\text{fission}}/\Sigma_{\text{fuel}}$

Number of fast neutrons produced by fission $\eta f\phi$

where η is $\nu\,\Sigma_{\text{fission}}/\Sigma_{\text{fuel}}$

Number of neutrons after fast fission enhancement $\varepsilon\eta f\phi$

Now in the thermal system the neutrons slow down through the resonance absorption region:

Number which escape resonance capture $p\varepsilon\eta f\phi = k_\infty\phi$

Number which escape leakage during slowing down $k_\infty\phi\exp(-B^2\tau)$

Having now arrived at thermal energies the neutrons diffuse until they again have a chance to be absorbed in fuel or are nonproductively absorbed in structure or poison material.

Number which escape leakage during diffusion $k_\infty\phi\exp(-B^2\tau)/(1+L^2B^2)$

Final thermal flux, $k_{\text{eff}}\phi$ $k_\infty\phi[\exp(-B^2\tau)/(1+L^2B^2)]$

$$(1.1)$$

where k_{eff} is called the neutron multiplication factor. The system is now critical if $k_{\text{eff}} = 1.0$ since the original neutron flux ϕ is not diminished after an average generation time. If k_{eff} is less than one, the neutron population dies away and the system is subcritical, while if k_{eff} is greater than one, the system is supercritical and the population continues to grow.

In fast reactors however there is little slowing down so there is very little resonance capture and p is close to unity and there is little leakage at this stage. The fast fission enhancement factor ε is of course not a useful concept when we are only concerned with fast neutrons. Thus for a fast system

$$k_{\text{eff}} \simeq \eta fp/(1+L^2B^2) \tag{1.2}$$

Although k_{eff} being unity implies criticality in the steady state, a fraction β of the neutrons, delayed neutrons, only arise some time later. Thus on the short term basis only $k_{\text{eff}}(1 - \beta)$ are immediately available. Thus the prompt criticality condition is $k_{\text{eff}}(1 - \beta) = 1.0$.

When $k_{\text{eff}}(1 - \beta) > 1.0$, the delayed neutrons have an insignificant effect since the system is critical on prompt neutrons alone. However, if $k_{\text{eff}}(1 - \beta) < 1.0$ and $k_{\text{eff}} = 1.0$, then the system is not critical until the delayed neutrons are produced; the delayed neutrons allow the control of the eventually critical system.

The cross sections Σ_{fuel}, Σ_{fission}, Σ_{a}, and the resonance escape probability p are all dependent on temperature. The fuel and other system temperatures are involved and their effect is given by temperature coefficients of reactivity α_i where

$$k_{\text{eff}} = k_{\text{eff0}} + \sum_i \alpha_i (T_i - T_{i0})$$

for each temperature T_i. Section 1.4 will refer to these effects in detail.

1.2.2 DIFFUSION EQUATION

The transient behavior of the neutron flux in a reactor core is represented by the diffusion equation (*3*):

$$\partial n/\partial t = \exp(-B^2\tau)k_\infty(1 - \beta)\Sigma_{\text{a}}\phi + \exp(-B^2\tau_{\text{D}})p \sum_i^N \lambda_i C_i + D\nabla^2\phi - \Sigma_{\text{a}}\phi \tag{1.3}$$

which expresses the fact that the rate of increase of neutrons is equal to the prompt and delayed neutron production reduced by leakage and absorption (*3*).

The delayed neutrons' concentrations are represented by:

$$\partial C_i/\partial t = (k_\infty\beta_i\Sigma_{\text{a}}\phi/p) - \lambda_i C_i, \qquad i = 1, 2, \ldots, N \tag{1.4}$$

which shows the radioactive decay of the precursors of the delayed neutrons.

Equation (1.3) is space, time, and energy dependent [$n(\mathbf{r}, t, E)$]. We could more accurately start with a set of equations each of which applies to a different energy. Here are diffusion equations for three discrete energy bands as an example:

Fast group (>1 MeV). Here fast neutrons are born, there is no resonance capture, but leakage occurs and the neutrons scatter down in energy to the next group

$$\partial n_{\text{F}}/\partial t = [k_\infty(1 - \beta)\Sigma_{\text{a}}\phi_{\text{T}}/p] + D_{\text{F}}\nabla^2\phi_{\text{F}} - \Sigma_{\text{F}}\phi_{\text{F}} \tag{1.5}$$

In the fast reactor Σ_{F} is small.

Epithermal group (1 keV–0.5 MeV). Here delayed neutrons are born, and further neutrons arrive from the fast group after energy degradation

$$\partial n_{\rm E}/\partial t = p\Sigma_{\rm F}\exp(-B^2[\tau - \tau_{\rm D}])\phi_{\rm F} + p\sum_{i=1}^{N}\lambda_i C_i - \Sigma_{\rm E}\phi_{\rm E} + D_{\rm E}\nabla^2\phi_{\rm E} \quad (1.6)$$

In the fast reactor $\Sigma_{\rm E}$ is even smaller than $\Sigma_{\rm F}$.

Thermal group (0.025 eV). Here all neutrons arrive after further energy degradation in a thermal reactor, although very few neutrons would reach this range in the fast reactor core

$$\partial n_{\rm T}/\partial t = \Sigma_{\rm E}\exp(-B^2\tau_{\rm D})\phi_{\rm E} + D_{\rm T}\nabla^2\phi_{\rm T} - \Sigma_a\phi_{\rm T} \quad (1.7)$$

These three equations may be summed to give Eq. (1.3) by defining a new diffusion coefficient D as:

$$D = D_{\rm T} + [p\exp(-B^2\tau)D_{\rm F}\nabla^2\phi_{\rm F}/\nabla^2\phi_{\rm T}] + [\exp(-B^2\tau_{\rm D})D_{\rm E}\nabla^2\phi_{\rm E}/\nabla^2\phi_{\rm T}] \quad (1.8)$$

$$n = n_{\rm F}p\exp(-B^2\tau) + n_{\rm E}\exp(-B^2\tau_{\rm D}) + n_{\rm T}$$

Thus the original single diffusion equation can be used to represent the single average group of neutrons. So long as D is chosen correctly the equation is more accurate than is, at first sight, apparent. Note that in a steady state, Eq. (1.3) reduces to the critical condition $k_{\rm eff} = 1.0$.

This reduction to a single group equation would also apply in the case of a fast reactor; however, one should note that it is an approximation that may be used in kinetic calculations while the accuracy would certainly not be adequate for steady-state physics calculations.

1.2.2.1 *Single-Point Kinetic Equations*[†]

Equation (1.3) is still space and time dependent even though the energy dependence can be averaged out by a judicious choice of the diffusion coefficient D. The main space dependent term is eliminated if the flux shape is assumed to remain constant as far as leakage is concerned during the kinetic calculation. The critical distribution derived from Fermi age theory (*3*) may be assumed:

$$\nabla^2\phi = -B^2\phi \quad (1.9)$$

where B^2 is the geometric buckling.

[†] See Glasstone and Edlund (*3*).

Then, with the following mathematical transformations

$$L^2 = D/\Sigma_a, \quad l_0 = 1/v\Sigma_a, \quad l^* = l_0/(1 + L^2B^2), \quad \phi_i = p\exp(-B^2\tau)C_i/l_0\Sigma_a,$$

and assuming that the Fermi ages τ and τ_D are equal and that the average neutron velocity is constant in time, we can further simplify Eq. (1.3).

In graphite-moderated systems typically $\tau = 350$ cm^2 and $\tau_D = 333$ cm^2 so that the above assumption is reasonable. In water-moderated systems where $\tau = 25$ cm^2 and $\tau_D = 14$ cm^2 the assumption is not so good. However in the fast reactor τ and τ_D are both very large and the assumption is excellent. Thus Eqs. (1.3) and (1.4) reduce to

$$\partial\phi/\partial t = [(\delta k - \beta k_{eff})\,\phi/l^*] + \sum_{i=1}^{N} \lambda_i\phi_i$$
$$\partial\phi_i/\partial t = [\beta_i k_{eff}\phi/l^*] - \lambda_i\phi_i, \quad i = 1, 2, \ldots, N \tag{1.10}$$

These equations are time dependent and ϕ may represent the total neutron flux in the system. In a fast system, which is relatively small, the approximation is good.

1.2.3 In-Hour Equation[†]

If we assume the separability of time and space dependence in the flux, for a step change of δk at $t = 0$, the flux and the precursor concentrations will eventually attain a steady period

$$\phi(E, t) = \phi_0 \exp(t\omega)$$
$$\phi_i(E, t) = \phi_{i0} \exp(t\omega) \tag{1.11}$$

After substitution in the single-point Eqs. (1.10) and cancelation of ϕ_{i0}/ϕ_0 and $\exp(t\omega)$, the in-hour equation is obtained:

$$\varrho = \delta k/k_{eff} = (l^*\omega/k_{eff}) + \sum_{i=1}^{N} \omega\beta_i/(\omega + \lambda_i) \tag{1.12}$$

This equation relates all the $N + 1$ time constants of transient ϕ and ϕ_i behavior for the reactivity level ϱ. Thus

$$\phi = \phi_0 \sum_{n=1}^{N+1} a_n \exp(t\omega_n), \quad \phi_i = \phi_{i0} \sum_{n=1}^{N+1} b_{in} \exp(t\omega_n)$$

For $\varrho > 1$ there is at least one dominant positive root ω_0 and the neutron flux increases with time. Figure 1.3 shows ω_0 as a function of ϱ.

[†] See Glasstone and Edlund (*3*).

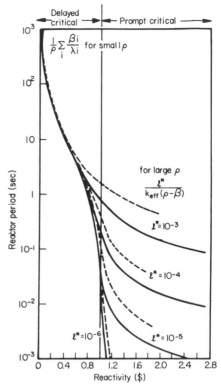

Fig. 1.3. Reactor stable period for ^{235}U thermal fission delayed neutron data (*12*)
Legend: — ^{235}U; – – – ^{239}Pu.

1.2.3.1 *Small Values of ϱ*

If ϱ is very small, then ω_0 is also small and less than the smallest λ_i. Thus

$$\varrho = (l^*\omega_0/k_{\text{eff}}) + \omega_0 \sum_{i=1}^{N} (\beta_i/\lambda_i) = l\omega_0 \qquad (1.13)$$

Thus the stable period, $\omega_0 = \varrho/l$, is inversely proportional to the mean effective neutron lifetime l where:

$$l = (l^*/k_{\text{eff}}) + \sum_{i=1}^{N} (\beta_i/\lambda_i) \qquad (1.14)$$

For a fast reactor l^* is between 10^{-6} and 10^{-8} and thus the terms of the right hand side of the equation for l are of the order of 10^{-6} or less and 40×10^{-3} so that l^* is not important here. For small reactivity changes then the delayed neutron characteristics are most important and

$$l \simeq \sum_{i=1}^{N} (\beta_i/\lambda_i) \qquad (1.15)$$

This is true also for thermal systems where $l^* \simeq 10^{-3}$.

1.2.3.2 *Large Values of* ϱ

If ϱ is very large then ω_0 is large and greater than the largest λ_i. Thus:

$$\varrho = (l^*\omega_0/k_{\text{eff}}) + \sum_{i=1}^{N} \beta_i \quad \text{so} \quad \omega_0 = (\varrho - \beta) k_{\text{eff}}/l^* \qquad (1.16)$$

Now the neutron lifetime is all important and the delayed neutrons have no effect beyond reducing the reactivity ϱ by β to $(\varrho - \beta)$.

1.2.4 DELAYED NEUTRON DATA

The decay constants of the delayed neutron groups λ_i do not differ markedly among fissionable isotopes, but the yields β do (4). A comparison of β for different fuels is shown in the accompanying tabulation. Thus an

Type of fission	Isotope	β
Fast	^{238}U	0.01560
Fast	^{239}Pu	0.00206
Fast	^{241}Pu	0.00530
Fast	^{232}U	0.02110
Thermal	^{235}U	0.00644
Thermal	^{233}U	0.00264

effective β must be calculated before reactor kinetics calculations can proceed. The value of β depends not only on the fuel in the system but on a large number of other system variables:

(a) The isotopic concentrations in the system and the fission rate in each isotope affect β. In a fast reactor ^{238}U may have 15% of the fissions.

(b) Delayed neutrons are slower than prompt neutrons and thus their importance with respect to leakage and fission is different (5). Their slowness reduces the effective β value in the fast core.

(c) The geometry of the system and the blanket fission rate effect β (6). In a fast system 10% of the fissions can occur in the blanket.

(d) Neutron hold-up in a reflector is sometimes represented by another delayed neutron group, although it is better represented by a modified neutron lifetime. This effect is particularly important in heavy water thermal systems but does not apply in a fast reactor system.

(e) Variation with burn-up also changes the balance as far as β is concerned. The plutonium is enhanced; thus the effective β decreases in a thermal system that may start with $\beta \simeq 0.0075$, but in fast cores where ^{239}Pu gives way to ^{241}Pu the effective β increases. A fast reactor typically has a β of about 0.0035.

1.2.4.1 *Number of Delayed Neutron Groups*

There is nothing magical about the usual six groups of delayed neutrons, other than the fact that the decay characteristics of a shut-down infinite system can be represented adequately by six exponentials and six yields. These do not vary greatly for thermal or fast fission except in the final value of β.

There has been some attempt to identify the groups with some predominant fission product but after ^{87}Br that corresponds to the first group, there has been no success.

TABLE 1.2

DELAYED NEUTRON DATA FOR FAST FISSION IN ^{239}Pu[a]

i	$10^3\beta$[b]	λ_i (sec^{-1})	$\tau_i = 1/\lambda_i$ (sec)	Half-life (sec)	Fission product
1	0.07828	0.0129	77.52	53.75	^{87}Br
2	0.5768	0.0311	32.16	22.29	^{137}I, ^{88}Br
3	0.4449	0.1335	7.49	5.19	
4	0.6757	0.3315	3.016	2.09	
5	0.2122	1.2623	0.792	0.549	
$N = 6$	0.0721	3.2083	0.312	0.216	

[a] See Keepin *et al.* (4).
[b] $\sum_{i=1}^{6} \beta_i = 2.06 \times 10^{-3}$.

Notice the mathematical progression in λ_i anyway. (See Table 1.2.) The progression of λ_i values (each approximately e times the previous one) arises from a mathematical matching or peeling of the decay curve rather than from actual fission product decay times.

In some problems where the number of equations makes the problem computationally large, say on an analog computer, or where the effect of delayed neutrons is small, then it is usual to take less than six groups.

A smaller number of groups can be arranged to represent the average behavior very well:

No groups ($N = 0$). One simply modifies the neutron lifetime in equation 1.10 so that

$$l = l^* + \sum_{i=1}^{6} (\beta_i/\lambda_i)$$

The equation then gives the correct stable period for small reactivity changes.

One group ($N = 1$). Define $\bar{\beta}$ and $\bar{\lambda}$ so that:

$$\bar{\beta} = \sum_{i=1}^{6} \beta_i, \qquad \bar{\beta}/\bar{\lambda} = \sum_{i=1}^{6} (\beta_i/\lambda_i) \tag{1.17}$$

The equation now satifies the high and low frequency response gain and phase (Fig. 1.4). See also Section 1.5.3 for a frequency response discussion.

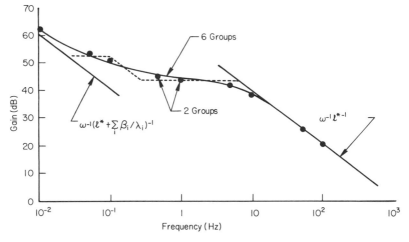

Fig. 1.4. Matching the frequency response of the neutron kinetics by few group data (*12*).

Two groups ($N = 2$). The smallest number of groups with which one can represent delayed neutron behavior adequately is two.

The relevant data can be obtained by matching particular transients, from the in-hour equation, or by matching the transfer function in more detail than for a single group. These methods are all discriminatory in that they relate best to particular transients, to steady periods, or to certain frequency disturbances, respectively. However they all give very adequate results.

Table 1.3 shows figures derived from a transfer function match for a plutonium-fueled reactor. The method is general and can be applied to any reactor system.

TABLE 1.3

TWO-GROUP DELAYED NEUTRON MATCH

j	$10^3 \beta_j{}^a$	λ_j (sec^{-1})
1	0.460	0.0260
2	1.524	0.2640

^{a} Note: $\sum_{j=1}^{2} \beta_j \neq \sum_{i=1}^{6} \beta_i$.

Figure 1.5 shows the same transient, a \$ 0.5 reactivity decrease as a step, represented by Eqs. (1.10), using different numbers of delayed neutron groups. It can be seen that while the modified effective neutron lifetime model at least gives the correct trend, the two-group representation is very adequate for transient calculations.

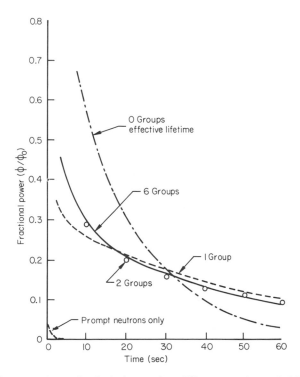

Fig. 1.5. A comparison of calculations using different numbers of delayed neutron groups (*12*).

1.3 Thermal Effects

The neutron population is a measure of the number of fissions that occur in the system, which is in turn a measure of the power produced. Thus the neutron concentration or flux is directly proportional to the power of the system. It is now necessary to understand how this power produces temperature changes that might effect the reactivity through temperature coefficients. This understanding can come from attempting to model the system in mathematical terms.

1.3.1 BASIC MATHEMATICAL MODEL

The transfer of heat through any material is governed by the heat balance equation:

$$\varrho c \, \partial T/\partial t = k \, \nabla^2 T + H \qquad (1.18)$$

where H is the heat production term. This, like Eq. (1.3), is also time and space dependent.

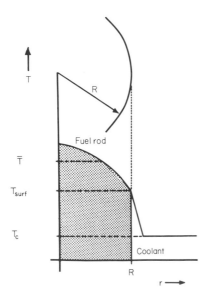

Fig. 1.6. Schematic representation of the thermal distribution through a cross section of a cylindrical fuel pin.

Considering a cylindrical fuel pin in which heat is produced (Fig. 1.6), Eq. (1.18) can be used to derive the average fuel temperature in the following way. In the radially symmetrical pin

$$\varrho c \, \partial T/\partial t = a\phi + k[(\partial^2 T/\partial r^2) + r^{-1}(\partial T/\partial r)] \qquad (1.19)$$

and by defining a volume-averaged fuel temperature and volume-averaged flux (or power) as

$$\bar{T} = (2\pi/A) \int_0^R Tr\, dr \quad \text{and} \quad \bar{\phi} = (2\pi/A) \int_0^R \phi r\, dr \quad (1.20)$$

where

$$A = 2\pi \int_0^R r\, dr$$

Equation (1.18) becomes

$$A\varrho c(d\bar{T}/dt) = a\bar{\phi}A + 2\pi k R(\partial T/\partial r)_{\text{at}R} \quad (1.21)$$

By now assuming that

$$(\partial T/\partial r)]_{\text{at}R} = (T_{\text{surf}} - T_{\text{c}})/\varDelta$$

we have

$$mc\, d\bar{T}/dt = a'\bar{\phi} - (2\pi R k/\varDelta)(T_{\text{surf}} - T_{\text{c}}) \quad (1.22)$$

and defining the heat-transfer coefficient $h = 2\pi R k/\varDelta$, the average fuel temperature is given by

$$mc\, d\bar{T}/dt = a'\bar{\phi} - h(T_{\text{surf}} - T_{\text{c}}) \quad (1.23)$$

Alternatively,

$$mc\, (d\bar{T}/dt) = a'\bar{\phi} - h'(\bar{T} - T_{\text{c}}) \quad (1.24)$$

Now the heat-transfer coefficient h' must include not only an allowance h for the film drop heat transfer between the surface temperature T_{surf} and the coolant temperature T_{c} as before, but also an allowance for conduction between the point of average fuel temperature $\bar{T}$ and the surface of the fuel. So

$$1/h' = (1/h) + (1/8\pi k) \quad (1.25)$$

This adjustment is made so that Eq. (1.24) enables $\bar{T}$ to be directly calculated whereas using Eq. (1.23) would have required an additional equation to obtain T_{surf} from $\bar{T}$ and T_{c}. Thus for the average fuel temperature, the equation becomes (Fig 1.7)

$$m_{\text{f}}c_{\text{f}}(dT_{\text{f}}/dt) = a\phi - h_{\text{f}}(T_{\text{f}} - T_{\text{c}}) \quad (1.26)$$

In a similar manner the average coolant temperature equation becomes

$$m_{\text{c}}c_{\text{c}}(dT_{\text{c}}/dt) = h_{\text{f}}(T_{\text{f}} - T_{\text{c}}) + h_{\text{s}}(T_{\text{s}} - T_{\text{c}}) - m_{\text{c}}c_{\text{c}}v_{\text{c}}(dT_{\text{c}}/dz) \quad (1.27)$$

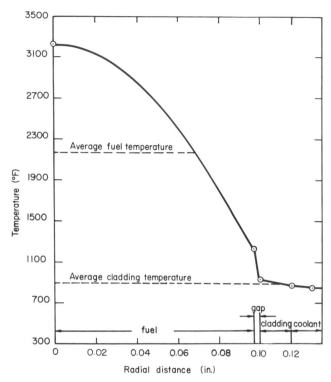

Fig. 1.7. Thermal distribution through a cross section of a cylindrical fuel pin and cladding.

where the last term accounts for the transfer of heat as the fluid moves down the channel with velocity v_c. Average temperatures of any other material such as structural components in the core can be established in an exactly similar manner; i.e.,

$$m_s c_s (dT_s/dt) = b\phi - h_s(T_s - T_c) \tag{1.28}$$

where in this case $b\phi$ accounts for any direct γ heating of the structure. Equation (1.27) accounts for heat deposited in the coolant from the fuel element $h_f(T_f - T_c)$ and from the structural material $h_s(T_s - T_c)$.

1.3.1.1 *Steady-State Temperatures*

With all time derivative terms set to zero, by adding the steady-state temperatures of this system (fuel, coolant, and structure) we can obtain

$$dT_c/dz = (a + b)\phi/m_c c_c v_c \tag{1.29}$$

$$T_c = T_{c_{in}} + [(a + b) \int_0^z \phi \, dz]/m_c c_c v_c \tag{1.30}$$

$$T_f = T_c + a\phi/h_f \tag{1.31}$$

$$T_s = T_c + b\phi/h_s \tag{1.32}$$

Thus one can obtain a complete distribution of temperatures as a function of z if the heat input distribution $\phi(z)$ and the coolant inlet temperature $T_{c_{in}}$ are known.

It is worth noting that the inlet temperature is effectively a base temperature for the system. Any increase in coolant inlet temperature is reflected by an identical increase in all the other temperatures in the steady state—if the power is retained constant. In practice, we shall see in the next section that, in the absence of external controls, the power ϕ will also change as a function of temperature because of feedback reactivity changes. However this change of power is a secondary effect.

1.3.1.2 *Time Constants*

All components within the reactor system have natural temperature time constants which are included in the model equations (1.26)–(1.28). They are:

Fuel

$$\tau_f = m_f c_f/h_f \tag{1.33}$$

For a graphite moderated thermal reactor with a Magnox fuel element, this time constant could be 5–10 sec; for a light water thermal system fueled with plates or small pins, the time constant is reduced to 1–5 sec; while for a fast reactor fuel pin, the time constant is 1 or 2 sec at the most. The fast reactor fuel therefore reacts very rapidly to power changes and the time which it takes to reach a higher temperature is reduced compared to the thermal systems.

Structural components

$$\tau_s = m_s c_s/h_s \tag{1.34}$$

For a graphite moderated system, the graphite itself might have time constants in the range of 10 to 15 min; however, in the fast reactor almost all structure reacts in times of less than a few seconds.

Coolant

$$\tau_c = L/v_c \tag{1.35}$$

The effective time constant here is the channel transit time which might be 0.5 sec for a light water thermal reactor but is about 0.1 sec for a sodium-cooled fast breeder.

Thus it can be seen that the fast reactor reacts to disturbances very much more rapidly than thermal systems in almost all respects. It is therefore less liable to some of the instabilities exhibited in the other systems, but it requires faster operating control and protective systems to ensure an acceptable response to reactivity and flow disturbances.

1.3.2 HEAT-TRANSFER COEFFICIENTS

The surface temperature drop to the coolant is calculated from an empirical heat-transfer coefficient that depends on the channel geometry, the fuel element and its cladding, and the coolant velocity and thermodynamic properties. There are a number of such correlations for different coolants and different ranges of operation.

Water. The Dittus–Boelter correlation (7a) is

$$N_u = 0.023(R_e)^{0.8}(P_r)^{0.33} \tag{1.36}$$

Sodium. The main correlations all follow a general equation of the form (7b)

$$N_u = A + 0.025(R_e)^{0.8}(P_r)^{0.8} \tag{1.37}$$

and in both cases the heat-transfer coefficient h can be calculated from the Nusselt number N_u, given by

$$N_u = hd_e/kA \tag{1.38}$$

These correlations usually result in the heat transfer being proportional to flow to the 0.8 power because of the Reynolds number dependence.

Internal heat-transfer coefficients are calculated from the conductivities combined as in Eq. (1.25). The main uncertainties occur in not knowing the condition of the fuel or the fuel/cladding gap.

Alternatively heat-transfer coefficients can be calculated from the steady-state temperature distributions for a given heat input [Eqs. (1.29)–(1.32)] if they are known from experimental measurements.

1.3.3 ELABORATION OF MODELS

Before considering the enlargement of the thermal model to take account of further spatial effects, there is one term which complicates the solution of the present set of Eqs. (1.26)–(1.28).

1.3.3.1 *The Term* $\partial T_c/\partial z$

The term $\partial T_c/\partial z$ in Eq. (1.27) complicates the solution because of the second implicit independent variable z. We can evade the complication by averaging the coolant temperature and assuming a mean gradient. This is good if the flux distribution is sinusoidal. Thus the usual approximation is

$$\partial \bar{T}_c/\partial z = (T_{c_{out}} - T_{c_{in}})/L = 2(\bar{T}_c - T_{c_{in}})/L \qquad (1.39)$$

where $\bar{T}_c$ is now a channel volume weighted mean.

However when temperatures are used to predict reactivity changes, whole reactor averages are required rather than component averages and these need to be weighted, according to perturbation theory, by the square of the flux ϕ (3). Thus

$$T_{av} = \int T\phi^2 \, dV / \int \phi^2 \, dV \qquad (1.40)$$

and for a sinusoidal flux this gives

$$\partial T_{av}/\partial z = 2.33 \, (T_{av} - T_{c_{in}})/L \qquad (1.41)$$

This is a more accurate value for feedback calculations and the average temperature here has a different meaning. The effect is a multiplicative constant, 1.16, which can easily be included in the model. However the fast reactor has a relatively flat flux and this weighting is less significant.

1.3.3.2 *Spatial Representation*

The present core thermal model is very coarse since it has only a single fuel and a single coolant temperature. The spatial representation can be improved by including more points in the fuel and the structure within the unit cell. This is done by adding further heat balance equations in each of which the temperature is the average of a particular volume.

$$m_{f1}c_{f1} \, dT_{f1}/dt = (1 - \gamma)(1 - \alpha)\phi - h_{f1}(T_{f1} - T_{f2})$$

$$m_{f2}c_{f2} \, dT_{f2}/dt = \gamma(1 - \alpha)\phi + h_{f1}(T_{f1} - T_{f2}) - h_{f2}(T_{f2} - T_{cd})$$

$$m_{cd}c_{cd} \, dT_{cd}/dt = h_{f2}(Tf_2 - Tc_d) - h_{cd}(T_{cd} - T_c) - R(T_{cd}^{*4} - T_s^{*4}) \qquad (1.42)$$

$$m_c c_c \, dT_c/dt = h_{cd}(T_{cd} - T_c) + h_s(T_s - T_c) - 2.33 m_c c_c v_c(T_c - T_{c_{in}})/L$$

$$m_s c_s \, dT_s/dt = \alpha\phi - h_s(T_s - T_c) + R(T_{cd}^{*4} - T_s^{*4})$$

Now we have two points in the fuel, one each in the cladding and coolant and one in the structural material.

Notice the radiative cooling term of the cladding. The problem is now nonlinear because of this term and will require more elaborate solution even in the steady state. However radiation terms are usually very small except in accident conditions; they may therefore be neglected in operational or near operational conditions.

Further axial representation and a better model for $\partial T_c/\partial z$ demands a set of such equations at a number of points of the coolant channel, linked through the coolant equation term $\partial T_c/\partial z$ by a finite difference representation. One such representation is the Fox–Goodwin equation (8a).

$$dT_{c_{n+1}}/dz + dT_{c_n}/dt = 2(T_{c_{n+1}} - T_{c_n})/\Delta z \tag{1.43}$$

This defines gradients and temperatures at successive points in the coolant channel (Fig. 1.8).

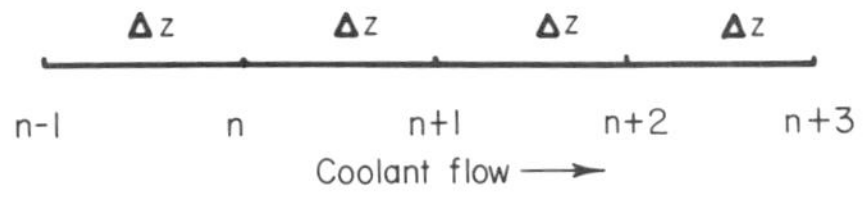

Fig. 1.8. Finite difference nomenclature along the reactor coolant channel.

With this model in steady state, neglecting the term $m_c c_c(\partial T_c/\partial t)$ and omitting the radiative terms,

$$T_{c_{n+1}} = T_{c_n} + (\Delta z/2m_c c_c v_c A')[h_{cd}(T_{cd_n} + T_{cd_{n+1}}) + h_s(T_{s_n} + T_{s_{n+1}})] - BT_{c_n} \tag{1.44}$$

where

$$A' = 1 + (\Delta z/2v_c)[(h_{cd}/m_c c_c) + (h_s/m_c c_c)] \quad \text{and} \quad B = (2A' - 2)/A' \tag{1.45}$$

thus defining the coolant temperatures at a point $n + 1$, in terms of heat input to that channel at that point $n + 1$, and upstream n.

Thus a set of such equations can be obtained at each axial point and for every channel whose representation is required. All of these equations depend on a knowledge of the inlet temperature and the power distribution for their solution.

1.3.3.3 *Special Effects*

Special effects, such as Wigner energy release $[A \exp(BT)(\partial T/\partial t)]$ in graphite structure systems or uranium oxidation $[a \exp(bT)]$ in metal-fueled systems, may have to be included in safety studies when adverse conditions are studied. However in sodium-cooled systems extra power production terms do not arise except when fuel is supposed to be ejected into the coolant channel. Then the coolant equation contains an extra heat production term. However, then it is also necessary to represent channel boiling (Section 1.3.4).

1.3.4 BOILING CHANNELS

1.3.4.1 *Two-Phase Boiling Model*[†]

If heat is input uniformly into a liquid coolant channel (Fig. 1.9), the temperatures of the fuel element surface T_{cd} and the coolant T_c increase as the coolant flows along the channel and the sequence of heating regions is as follows:

(a) *Convective* heat transfer region, in which the heat transferred q is given by

$$q = h(T_{cd} - T_c) \tag{1.46}$$

T_c and T_{cd} increase linearly and T_{cd} reaches a value just above the coolant saturation temperature.

(b) *Highly subcooled* boiling region, in which heat transfer is mainly convective. Bubbles form and collapse on the element surface and T_c increases.

(c) *Slightly subcooled* boiling region, in which heat transfer is partly by convection and partly by a boiling mechanism. Bubbles persist and are swept into the stream only to diminish in size as they move along the channel. The value of T_c increases until the saturation value is reached.

(d) *Bulk-boiling* region, in which the heat transfer q is given by

$$q = \eta(T_{cd} - T_{sat})^4 \tag{1.47}$$

Steam bubbles grow in the stream and T_c remains constant as all the heat goes to convert liquid into vapor.

(e) *Superheated* region after the liquid is fully converted to vapor and the vapor temperature T_c and the surface temperature T_{cd} now both rise again.

A pressurized water reactor (PWR) channel terminates in the slightly subcooled region whereas a boiling water reactor (BWR) channel extends into the bulk-boiling region. In a steam generator, a section will also include the superheated region.

[†] See Tong (*8b*).

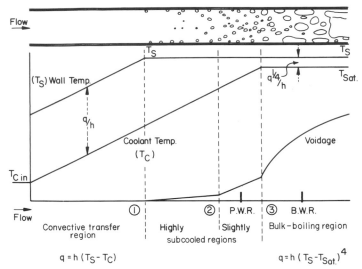

Fig. 1.9. Thermal effects within a boiling channel.

The mathematical representation of this two-phase situation uses the three conservation equations of energy, mass, and momentum to define the void fraction, the vapor velocity, and the pressure.

Energy

$$\frac{\partial}{\partial z}\left(\varrho_l[1-f]v_l h_l + \varrho_v f v_v h_v\right) + \frac{\partial}{\partial t}\left(\varrho_l[1-f]h_l + \varrho_v h_v f\right) = b\phi + \frac{\partial p}{\partial t} \pm G_1$$

(1.48)

Mass

$$\frac{\partial}{\partial z}\left(\varrho_l[1-f]v_l + \varrho_v f v_v\right) + \frac{\partial}{\partial t}\left(\varrho_l[1-f] + \varrho_v f\right) = 0 \qquad (1.49)$$

Momentum

$$\frac{\partial}{\partial z}\left(\varrho_l[1-f]v_l^2 + \varrho_v f v_v^2\right) + \frac{\partial}{\partial t}\left(\varrho_l[1-f]v_l + \varrho_v f v_v\right)$$

$$= -\frac{\partial p}{\partial z} - \frac{2\tau}{d_e}\varrho_l v_l^2 \pm G_2 \qquad (1.50)$$

However these equations presuppose some knowledge of the ratio of the vapor velocity to the liquid velocity v_v/v_l, the slip ratio, for which experimental correlations exist. The equations also assume that the two phases are in thermal equilibrium with each other. This is usually a good assumption except in very severe transients when the vapor and liquid might have to be considered separately. Note that the densities (ϱ_l, ϱ_v) and enthalpies (h_l, h_v) are all pressure dependent and the calculation is now very complex.

This model applies well to water and it may be used for water-cooled thermal reactors or in steam generator modeling. However sodium vaporizes at high superheats and the two-phase situation is over very rapidly, resulting in almost immediate bulk boiling (see Section 1.3.4.3).

1.3.4.2 *Boiling Boundaries*

The boundaries between different regions of the boiling channel are specified by experimental correlations. They are required in modeling because we need to know where the heat-transfer mechanism changes from convective to bulk-boiling and where voiding occurs.

In a steam generator problem, it is necessary to integrate the heat transfer correctly and to obtain accurate time constants. The latter depend entirely on the heat-transfer coefficients (see Section 1.3.1.2). In a water-cooled core channel it is necessary to know the total core voidage, because its reactivity effect may be large (due to a removal of moderator) and fast (due to flashing or collapse of vapor voids).

It is possible, knowing the position of the boundary between the highly and slightly subcooled regions, to simplify the channel representation in a PWR. Here it is assumed that convective heat transfer exists below the boundary and boiling transfer exists above it. The equations for the mixed enthalpy of the fluid are used, and another experimental correlation is used to predict total subcooled voidage in the channel. The analysis for a channel that includes subcooled nucleate boiling and the experimental correlations used to predict boiling boundaries and voidage have been given by Bowring (9) for water systems operating in certain temperature ranges. If however a channel includes bulk boiling, the small subcooled boiling range may be insignificant and the only boundary of interest is that marked by the saturation temperature or a certain superheat above this saturation temperature.

1.3.4.3 *Annular Slug Boiling Model*[†]

In sodium, boiling exhibits quite different characteristics than in water and stable boiling does not appear possible at reactor flow rates. The specific volume of the vapor at low pressure is high and it requires very high

[†] See Tilbrook and Macrae (10).

velocities for stability; higher than can be provided by available pump designs.

So in a sodium cooled system, inadvertent boiling is presumed to lead rapidly to a *bulk-boiling* region in which an annular flow representation is the best model. In this model the sodium vapor expands against the inertia of sodium liquid slugs above and below it. The value of T_c is constant at the saturation value because all the heat goes into converting more liquid into vapor. The bubble can extend outside the core into a region of significant condensation (Fig. 1.10).

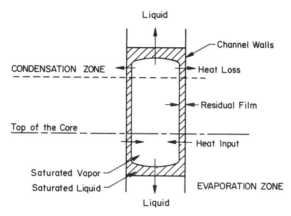

Fig. 1.10. A model for sodium vaporization in a coolant channel (*10*).

The mathematical representation uses the conservation equations of mass and energy for the saturated region to define the size of the bubble, the velocity of the interface between vapor and liquid, and the pressure within the bubble. This model has no need of special boundary representations between boiling and nonboiling regions, but there are considerable uncertainties. The uncertainties include:

(a) The superheat value at which boiling occurs, which may range as high as 500°F but is expected to be as low as 30–50°F in a reactor environment.

(b) Heat transfer data and the conditions under which a liquid film is retained on the fuel-pin cladding adjacent to the bubble.

(c) Condensation effects within a real system.

The condensation that the bubble experiences outside the core is most important since this defines the mode of sodium-vapor bubble collapse, which allows the sodium liquid slugs to reenter the core. This gives rise to presumed chugging motion (see also Section 4.4.2) in which the vapor bubble alternately grows and collapses until fuel failure results from the reduced heat transfer.

However boiling within a sodium-cooled breeder is an accident condition and may give rise to rapid fuel failure. Thus, in a design model of the core, sodium boiling would not be included and the design would almost certainly have considerable margins before boiling might be initiated. For a 35 psi cover gas system the sodium might boil at 1850°F whereas the highest coolant temperature would be approximately 1200°F, leaving a temperature margin of over 600°F.

1.4 Feedback Effects

It has been shown in preceding sections that one can predict the power generation level from the reactivity, and that temperatures can be predicted from this power level. It has been noted that the temperatures affect the reactivity through variations in the cross sections which determine the neutron behavior. Thus a complete feedback path exists (Fig. 1.11). It is necessary to know how reactivity depends on temperatures and how this dependence may be affected by the reactor and plant design.

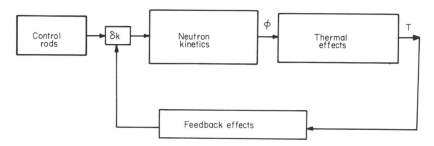

Fig. 1.11. A basic reactor feedback loop.

1.4.1 COEFFICIENTS OF REACTIVITY

The various system temperatures T_i affect the cross sections and escape factors which go to make up the multiplication factor k_{eff}.

$$\delta k = \delta k_0 + \left[\sum_i \alpha_i(T_i) \, (T_i - T_{i0}) \right] \tag{1.51}$$

The total feedback may be separated into its component effects as shown in the following discussion.

1.4.1.1 *Doppler Coefficient*

As the fuel temperature increases, the motion of the fertile atomic nuclei increases and the width of the absorption resonances increases (although the height decreases to keep the area constant). This occurs mainly below 20 keV. The thermal reactor resonance escape factor p is reduced and, in combination with the flux distribution, this results in a change in reaction rates giving a decrease in reactivity [Eq. (1.1)]. Thus, for a fuel temperature increase, the reactivity decreases so that α_{fuel} is negative.

However the opposite happens for ^{239}Pu because, on absorption, fission may occur in the plutonium and the reactivity increases. Here α_{fuel} may be positive. Thus the actual Doppler coefficient is determined by a competition between the two effects and it depends on the isotopic concentrations of ^{239}Pu and ^{238}U and their relative proximities. The plutonium effect is generally small.

The usual calculated values are approximately 10^{-5} $\delta k/k$ and are designed to be negative. The coefficient is the main safety parameter of a design as we shall see in later sections.

The Doppler coefficient (α_{fuel} or dk_{eff}/dT) in current liquid-metal fast breeder reactor designs varies as T^{-1}, so the Doppler coefficient is very often quoted in terms of the value of $T\ dk_{\text{eff}}/dT$, the Doppler constant.

1.4.1.2 *Moderator Coefficient in a Thermal Reactor*

In a thermal reactor, ^{238}U is a $1/v$ absorber in which the absorption cross section decreases as the inverse of the square root of the energy of the neutrons. However the absorption in ^{235}U falls off faster than this and the absorption for ^{239}Pu falls off more slowly than this as the energy increases (Fig. 1.12).

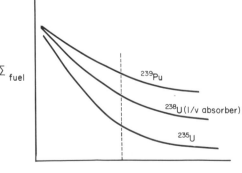

Energy (thermal)

Fig. 1.12. The energy dependence of the fuel absorption cross section for different isotopes (thermal system).

Thus, in a plutonium dominated system, the coefficient is usually positive but is very small in magnitude, whereas in a ^{235}U dominated system the coefficient is negative, but any value can result from this competition process.

However, the temperature of the system also has other effects which cannot be neglected: as the temperature and the energy E increases, the diffusion length increases, giving more leakage in small cores; the relative absorption in the fuel ($f = \Sigma_{\mathrm{fuel}}/\Sigma_{\mathrm{a}}$) increases for fine structure changes and the control rod effectiveness increases. These changes give, respectively, negative, positive, and negative contributions to a system temperature coefficient.

Thus the final moderator coefficient is very much a matter of balance and can either be positive or negative. It depends critically on the configuration and makeup of the design.

1.4.1.3 *Sodium Void Coefficient or Density Coefficient in a Fast Reactor*

This coefficient is very similar to the thermal density coefficient, but the emphasis is slightly different. The effects can be summarized in terms of the following.

(a) *Absorption variation.* The sodium removal reduces nonproductive absorption and relatively more neutrons are productively absorbed in the fuel. This is a positive effect.

(b) *Leakage changes.* As the diffusion length increases it leads to a greater leakage and gives a negative contribution to the coefficient.

(c) *Moderation changes.* There is less moderation and thus the fractional loss in energy per collision (the energy decrement) reduces and so the spectrum hardens. Thus the number of neutrons per fission (ν) increases and this too gives a positive contribution (Table 1.4).

(d) *Self-Shielding.* Self-shielding decreases giving a positive effect.

TABLE 1.4

COMPARATIVE ABSORPTION IN FAST AND THERMAL FLUXES

Isotope	σ_{a} (barns)	
	Fast	Thermal
^{239}Pu	2.36	10^3
Fe	0.01	2.6
^{135}Xe	3	$3 \cdot 10^6$
^{235}U	—	700

Thus the final sodium void coefficient is a matter of balance among these effects and the final value can be positive or negative. Because the effects are position dependent, the leakage being the dominating effect on the outer edges of the core and the moderation effect dominating at the core center, the sodium void coefficient is also very position dependent. Figure 1.13 shows the spatial variation of the sodium void coefficient and its components as a function of the core radius.

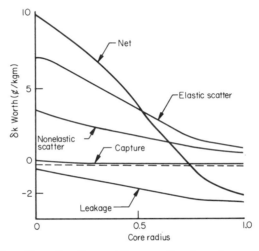

Fig. 1.13. The effect of sodium voiding as a function of core position (*11*).

1.4.2 EFFECTS OF CORE DESIGN[†]

1.4.2.1 *Spectrum*

Changes in core configuration or core materials all have an effect on the neutron spectrum (Fig. 1.14). It is instructive to note the differences between a light water thermal system and the current LMFBR designs.

(a) Sodium is substituted for the water coolant. This decreases the moderation and hardens the spectrum, thus leading to less parasitic and nonparasitic absorption effectiveness (Table 1.4).

(b) Stainless steel cladding is used instead of Zircaloy. This can be done, since parasitic absorptions are less important in the harder spectrum and absorption in steel is not significant. The steel also allows the use of higher cladding temperatures with consequent improved efficiencies.

[†] See Driscoll (*11*).

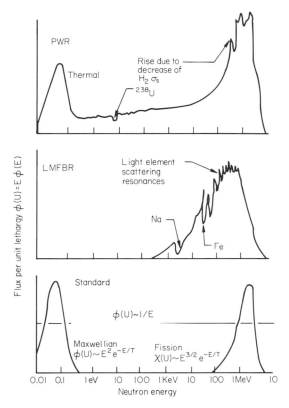

Fig. 1.14. Neutron spectra for PWR and LMFBR compared to standard distributions (*11*).

(c) The fissile enrichment is increased from about 3 to 15%. More enrichment is required to cope with increased inelastic scatter due to the harder spectrum.

(d) The volumetric proportions of fuel to coolant and structure are retained at the same levels (about 35:45:20) approximately, although, if anything, the LMFBR is a tighter lattice.

Thus the LMFBR has a harder spectrum, almost entirely above 100 eV, whereas the light water thermal system has a considerable thermal peak (Fig. 1.14).

The breeding ratio calculated from a static neutron balance in a clean core is the ratio of fertile captures to fissile absorption:

$$b \simeq (\eta_{239} - 1)\frac{\Sigma_f^{239}}{\Sigma_a^{239}} + \eta_{238}(1 - 1/\nu_{238})\frac{\Sigma_f^{238}}{\Sigma_a^{239}} - \frac{\Sigma_{ap}}{\Sigma_a^{239}} - \frac{\Sigma_B}{\Sigma_a^{239}} \qquad (1.52)$$

or approximately

$$b = \eta_{239} - 1 \tag{1.53}$$

where η_{239} is the number of neutrons produced per neutron absorbed.
In the hardest possible spectrum, η_{239} is highest and thus the harder the
spectrum the better the breeding ratio. Table 1.5 shows that breeding is
impossible with plutonium in a thermal environment, whereas it is very
good in a fast spectrum.

TABLE 1.5

NEUTRONS PRODUCED PER FISSION IN FAST AND THERMAL FISSION[a]

	$\eta = \nu\, \sigma_f/\sigma_a$	
Isotope	Thermal fission (PWR)	Fast fission (LMFBR)
^{233}U	2.28	2.64
^{235}U	2.05	2.46
^{239}Pu	1.96	3.03

[a] Minimum for criticality: $\eta = 1$; minimum for breeding: $\eta = 2$.

1.4.2.2 *Safety*

Very qualitatively then, safety can be said to penalize the breeding ratio.
If one designed for a large Doppler coefficient by having a softer spectrum
with more resonances below 10 keV playing a significant role, then the
breeding ratio would decrease.

Fortunately, however, other considerations also demand a softer spectrum. The ceramic fuel required to increase the fuel lifetime, the coolant
required to remove heat, the cladding for support and containment, and
the fertile material in the core to reduce the reactivity swing during burn-
up—all these degrade the spectrum and give rise to a reasonable Doppler
coefficient anyway.

However, one would like to soften the spectrum a little more to increase
the Doppler coefficient and gain an extra safety margin. This has to be done
while keeping strict account of the resultant breeding ratio. If there were
a particular need to enhance the Doppler feedback, then one *could* soften
the spectrum and improve the Doppler coefficient by:

(a) Choosing a low molecular-weight fuel with moderating atoms. Metal fuel gives the smallest Doppler coefficient and this is improved successively by changing to carbide, nitride, and then oxide fuels.

(b) Adding beryllium oxide to the core.

(c) Using boron and not tantalum rods for control, although then more boron would be required for the control required.

(d) Using a relatively low temperature and using the $1/T$ dependence of the Doppler coefficient. This solution is not available in a power producing system.

(e) Making sure that the plutonium and uranium isotopes are intimately mixed in the fuel. The point here is to keep the uranium that gives the negative contribution to the Doppler coefficient in close association with the plutonium where the neutrons are produced by fission and where, otherwise, only a positive Doppler coefficient would result. In fact, because the plutonium and uranium tend to migrate away from each other at high burn-ups, the Doppler tends to be delayed as a function of burn-up and could also be reduced if the separation were sufficient to cause spectrum changes.

The other coefficient of great interest in the sodium-cooled system is the sodium void coefficient. This too can be affected by the design.

Figure 1.15 shows the neutron worth (adjoint flux) as a function of energy plotted against the background of a fast spectrum. An increase in

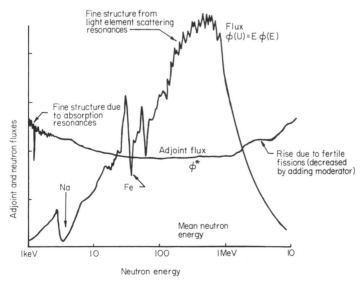

Fig. 1.15. The adjoint flux.

the mean neutron energy clearly results in an increase in reactivity. This is the main effect of voiding the center of a sodium-cooled core where leakage has very little effect, so that design changes seek to alleviate this effect by flattening out the worth curve or by moving the mean neutron energy lower to a positive slope of the worth curve.

Several design choices will accomplish this effect:

(a) Add a nonvoidable moderator such as beryllium oxide.

(b) Use NaK rather than sodium in the first place. It is a poorer moderator and thus its removal makes less difference to the neutron energy (but the use of NaK would also reduce the Doppler coefficient).

(c) Use dilute fuel to soften the spectrum as for the Doppler effect above.

(d) Use clean ^{239}Pu with no higher isotopes which have a greater worth in neutrons produced per fission.

(e) Use ^{233}U as the fissile species and so flatten the worth curve above 1 MeV.

(f) Use a higher core pressure to maintain a higher vapor density after voiding with a consequent slightly softer spectrum.

(g) Increase the sodium inventory in the core to make the spectrum much softer initially.

Then one can also use design to increase the leakage effect:

(h) Use a pancake or modular core to increase the leakage.

(i) Power flatten and decrease the core size and so increase the effect of leakage. This would not help at the center where the power is of course always flat.

However voiding effects are not necessarily all important; time effects may be more important in particular accidents and, of course, the avoidance of accidents in the first place, by providing a reliable system, is the primary objective of safety engineering.

1.4.2.3 *Poison Changes*

It is convenient first to outline the behavior of a poison that the fast reactor does not experience. In large thermal systems xenon produced from ^{135}Te by way of ^{135}I is now a very important poison. Because it is a byproduct of a fission product, it is flux-induced rather than temperature-induced. It is not an important poison in the fast reactor system.

The effect of xenon is complicated by its production from iodine and its destruction by a combination of neutron absorption and natural decay to ^{135}Cs and by the various time delays involved. The production and destruction is illustrated in Fig. 1.16.

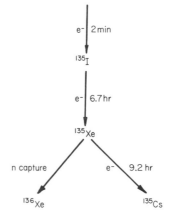

Fig. 1.16. Xenon production and destruction chain.

The relevant equations descriding the production and decay of iodine and production and destruction of xenon are:

$$\partial I/\partial t = \sigma_X \phi - \lambda_I I \tag{1.54}$$

$$\partial X/\partial t - \lambda_I I - \lambda_X X - \sigma_X \phi X \tag{1.55}$$

After shut-down, the xenon concentration does not see the lack of production of tellurium for a time delay of several hours due to the 6.7-hr iodine decay time. However it does see an immediate reduction in removal by neutron capture; thus the xenon concentration grows. This increase in xenon concentration complicates the subsequent start-up of some small thermal reactors due to the poison increase.

Further, xenon poisoning could produce spatial instabilities in the very large thermal systems, as some regions of the core could see poison changes on different time scales from others. This effect is corrected by regional control systems.

In the fast spectrum, xenon is not important since its parasitic absorption is very low (Table 1.4) and the fast cores are too small for spatial instabilities. However, see the thorium cycle reactivity change in the next section, as it is a similar though opposite effect.

Other fast reactor absorbers are given in Table 1.6.

1.4.2.4 *Dimensional Changes*

In fast reactors which have very small cores, the bowing or buckling of fuel elements due to temperature changes may shift fuel in or out of more reactive regions with consequent increases or decreases in reactivity.

TABLE 1.6

FAST REACTOR ABSORPTION CROSS SECTIONS FOR REACTOR MATERIALS

Material	σ_a (barns)	Use in reactor
^{238}U	0.35	Fuel
^{239}Pu	2.36	Fuel
Fe	0.01	Structural material
Na	0.0016	Coolant
B (natural)	0.54	Control
^{10}B	2.25	Control absorbers[a]
Ta	0.50	Control absorbers
^{6}Li	0.029	Coolant poison
^{7}Li	0.0000023	—

[a] There are problems associated with helium production in the rods and rod burn-up problems.

EBR-I exhibited both a prompt inward bowing that produced a positive reactivity change and a delayed outward bowing that produced a negative reactivity change (see Section 2.5.5). However, most power reactor designs would seek to keep such changes to an absolute minimum by employing a core restraint system, either by physically squeezing the core fuel assemblies together as in the Fermi reactor or by using a thermal restraint which uses the expansion of the core itself to tighten the core assemblies against fixed supports.

1.4.2.5 *Burn-Up Effects*[†]

As the fuel burn-up proceeds, the fuel composition and the control effectiveness changes and thus the reactivity balance of the system alters. These changes are also feedback effects, although the time scale is very large. Because these changes are long-term, they have no immediate effect on stability and so they are generally omitted from transient calculations. They are, however, accounted for by performing safety evaluations at several times during the burn-up cycle, particularly at start-of-life, during the equilibrium cycle, and at end-of-life.

We can separate out effects due to fuel and control changes:

[†] See Driscoll (*11*).

Fuel. (a) The reactivity effect of the core fissile concentration decrease can be expressed as

$$\frac{\delta k}{k} = -\frac{1}{2} \frac{\delta M}{M} = -\frac{1}{2} \frac{B \cdot 10^{-6}}{\varepsilon} \frac{1 + \alpha}{1 + \delta} (1 - b_i) \qquad (1.56)$$

If the internal breeding ratio b_i of the core is 0.9 and the fractional enrichment ε in the core is 0.16, then at a burn-up B of 50,000 MWD/tonne this reactivity change is -0.015 or about \$ 3 to \$ 4 negative. (Here the ratio of fissile captures to fissions α is taken to be 0.15 and the ratio of fertile to fissile fissions δ is assumed to be 0.2.)

(b) Fission product build-up can be expressed as

$$\frac{\delta k}{k} = -\frac{2}{5} \frac{B \cdot 10^{-6}}{\varepsilon} \frac{1 + \alpha}{1 + \delta} \frac{\sigma_{a_{fp}}}{\sigma_{a_{239}}} \qquad (1.57)$$

which includes the competition of absorption cross sections in the fission products $\sigma_{a_{fp}}$ and in the plutonium $\sigma_{a_{239}}$. This reactivity change is usually about \$ 2–\$ 3 negative.

(c) In ^{233}U/^{232}Th cycle systems, the reactivity effect is a swing due to the hold-up of the intermediate ^{233}Pa. There is a delay of about 30 days before a ^{232}Th capture results in the production of an atom of ^{233}U. This leads to a decrease in reactivity at the start of irradiation, with a consequent increase of reactivity after shut-down due to the decay of the ^{233}Pa. It is similar to a reverse xenon poisoning effect (see Section 1.4.2.3). The reactivity swing is given by

$$\frac{\delta k}{k} = \frac{1}{2} \frac{\phi}{\varepsilon} \frac{\sigma_{c_{232}}}{\lambda_{233}} \qquad (1.58)$$

where the protactinium decay constant λ_{233} is taken as $2.9 \cdot 10^{-7}$ per sec and the ^{232}Th capture cross section $\sigma_{c_{232}}$ is about 0.4 b. Thus, for a fast flux of $2 \cdot 10^{15}$ n/cm^2-sec, the reactivity swing is 0.008 or \$ 2.5.

Control assembly. (a) The ^{10}B absorber burns out at an appreciable rate of approximately one percent per month for an in-core rod. Therefore the rods must be replaced every year or two.

(b) The ^{10}B absorber captures by a (n, α) reaction and the helium gas inside a boron carbide rod leads to a pressure build-up inside the control assembly cladding, which also leads to a need to replace the rods.

(c) Fuel management compensation leads to possible multizone refueling schemes to reduce the reactivity swing by a factor of approximately 2. Such multizone refueling schemes of course lead to additional shut-down time.

1.4.2.6 *Control Effects*

Every power plant has an added control system. These systems may have small or large amounts of automatic functions. These functions might include a control of the reactor flow or reactivity balance as a result of temperature monitoring. For example, it might be required to maintain a constant or near constant outlet temperature. Thus this additional control function is also a feedback that needs consideration during any safety evaluation.

Such feedbacks are entirely dependent on the design of both the plant and its control system. However, control system criteria for the design will ensure that such feedbacks contribute to the stability of the system rather than detract from it. Indeed they have the ability to make the system stable despite any adverse inherent feedback loops there might be.

The next section will make further reference to control and protective system feedbacks in the reactor system closed loops.

1.5 System Behavior

Having now seen that the reactor system consists of a closed loop, involving the reactivity, power generation, temperatures, and back to the reactivity, and having seen that there are other feedbacks which operate with different time scales and from different initiators (temperature, flux, fuel composition, and control), it is necessary to see how the whole picture fits together for particular reactor systems.

The safety engineer is interested in setting up a complete model of the system in order to assess the various interactions within the model for their safety in the face of expected and unexpected external and internal disturbances. By modeling a fast reactor system and disturbing that model with a flow change which might arise, such as a pump coast-down, he can calculate resulting temperatures to determine whether the behavior is acceptable or not.

This section deals primarily with the principles of interacting variables. The next chapter treats specific disturbances to fast reactors cooled by gas, steam, and sodium but here we are concerned with seeing how the safety engineer deduces or calculates what the results of disturbances might be.

1.5.1 CLOSED-LOOP MODELS

Using the coefficients of reactivity we have the final equation for the system representation, the feedback equation

$$\delta k = \delta k_0 + \sum_i \alpha_i (T_i - T_{i0}) + \alpha_v \, \delta V + \alpha_B B \tag{1.59}$$

were temperature, voiding and burn-up feedbacks are included.

This equation, together with the kinetics equations and the thermal model equations, forms the complete representation of the reactor. The interdependence of these equations can be seen from closed-loop model diagrams.

1.5.1.1 *Reactor Cycle Types*

Figure 1.17 shows the basic blocks of the closed loop—the kinetics, the thermal representation, and the feedback from thermal effects. An added xenon feedback is included to indicate how the thermal reactor xenon poisoning arises directly from flux changes. (Note also that the delayed neutrons contribute to the calculation of the flux only through the kinetics block.)

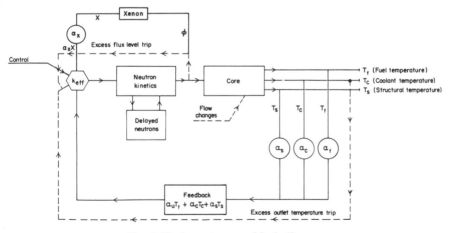

Fig. 1.17. A reactor core block diagram.

Further feedback lines can be added to this basic reactor core closed-loop model. They might arise from the control and protective system, and the figure shows the flux and excess outlet temperature trip lines as indicative of this type of feedback. External influences also play their part; here control and coolant flow changes might disturb the state of the system.

Indeed as we have seen, the coolant inlet temperature is the reference temperature for the system.

This basic feedback loop is added to in different ways for each different reactor type.

a. *Boiling water reactor (BWR)*. Figure 1.18 now shows the closed-loop model for the BWR. The circuit between coolant outlet and inlet is closed by a primary loop which includes the steam drum. The time delay between outlet and the drum is shown as τ_3 and the subsequent delay to the inlet plenum is denoted by τ_4. Conditions in the steam drum depend on a representation of how much boiling is occurring in the core and what the outlet quality is and also on what steam load is extracted from the drum to

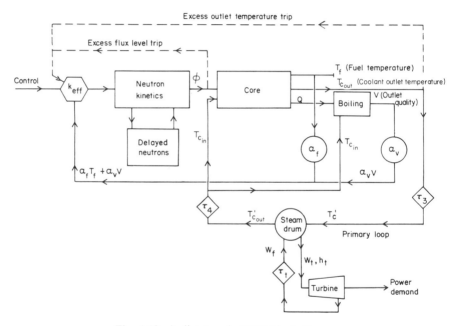

Fig. 1.18. A direct cycle BWR block diagram.

drive the turbine. Thus this model also requires a turbine representation and its return make-up lines to the steam drum. Notice also that the boiling in the core contributes a feedback reactivity to k_{eff}. The closed loop is an example of a direct cycle system.

b. *Pressurized water reactor (PWR)*. The circuit (Fig. 1.19) is very slightly altered by the addition of another secondary circuit as the PWR does not have boiling and must generate the steam for the turbine in a heat exchanger

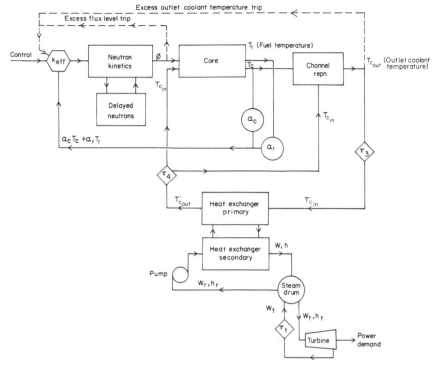

Fig. 1.19. An indirect cycle PWR block diagram.

(steam generator). It is an example of an indirect cycle system in which the steam drum and turbine are in the secondary cycle.

c. *Liquid metal fast breeder reactor* (*LMFBR*). Figure 1.20 shows that yet another circuit has been added between the secondary of the heat exchanger and the steam generator. This is an insulator between the primary radioactive sodium coolant and the water in the steam cycle. Now no xenon feedback exists.

External plant items in these loops will all be represented by thermal balance equations in as much complexity as the particular problem requires. The representation of the steam generator is difficult outside the normal operating range, because the boiling boundaries can shift so rapidly with steam collapse and water flashing into steam. The turbine, too, is particularly difficult since its transient behavior is largely unknown. Delays in the circuits must be included in the representations since they have a critical bearing on the stability of the system and these delays are variable, depending as they do on the circuit flows.

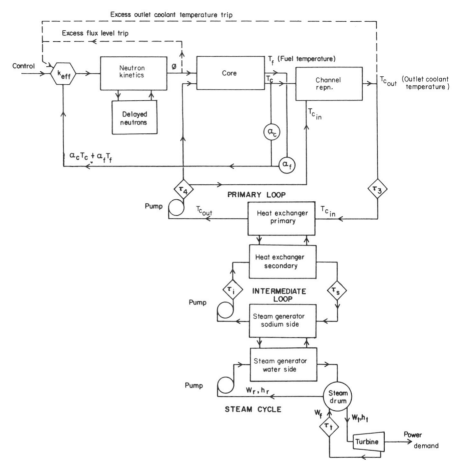

Fig. 1.20. A LMFBR block diagram in which an intermediate loop is added to the indirect cycle system.

It is possible to simplify models, depending on the investigation to be made. If, for example, a rapid reactor core variation is the phenomenon under consideration, it will be adequate to assume that the inlet temperature is constant (as it must be for a circuit time). This will allow the secondary and tertiary circuits to be neglected. If a secondary flow perturbation is to be investigated for its effect on the core, then it will be adequate in some circumstances to assume constant steam generator boundary conditions (in the LMFBR) and so obviate the need for representing the turbine and its feedwater line.

1.5.1.2 *Computation*

Having now constructed a mathematical model of the system and knowing the interaction between its relevant variables, we can calculate the system behavior resulting from certain disturbances.

These disturbances may be operational (the closure of a valve restricting steam flow or the insertion of rods to obtain a reactivity reduction), or they may be faults (pump failure, a burst steam main, or an inadvertent withdrawal of control). The model may be suitable for investigating each of them, or it may have to be modified for each set of circumstances. One might have to add a representation of the pumps in more detail if their behavior off their normal characteristic is required.

The calculations are performed either on an analog or a digital computer. The analog is especially convenient for time-dependent differential equations but the sheer size and nonlinearity of the calculation may demand the use of a digital machine. Presently it is customary for a reactor designer to have several standard codes that he may use as tools in the representation of certain types of transient—using one with a detailed fuel model, for example, to calculate the response to a reactivity transient and one with a detailed pump and circuit representation to calculate flow transient behavior. These are usually digital codes (see the Appendix).

Analog machines are in use for stability investigations and for control simulations, their range having been extended by the addition of some digital facilities, making them into hybrid computers with a combination of digital and analog functions.

1.5.2 Transients

Chapter 2 investigates the system transient behavior in detail for various fast reactors; here, we are interested in the reasons for performing the dynamic analyses rather than in the details of the particular transients themselves.

It is useful to relate fast reactor transients to light water thermal systems which have a wider variety of behavioral modes.

1.5.2.1 *Load variations*

Figure 1.21 illustrates the behavior of a PWR following a load change (*12*). The sequence of interactions is as follows:

(a) The turbine stop-valve is closed reducing the steam to the turbine.

(b) As a result of the closure, the secondary pressure increases in the steam generator and steam drum (see Fig. 1.19).

(c) The saturation temperature rises with the pressure.

(d) Heat transfer into the steam generator is inhibited by this rise of secondary temperature.

(e) The primary coolant temperature rises as less heat is being removed.

(f) As the coolant temperature coefficient of reactivity is negative, the neutron power reduces.

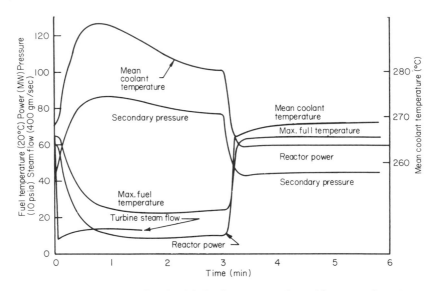

Fig. 1.21. A load change in a load-following system. The turbine steam flow demand is varied from 100% to 10% in 3 sec and then, after 3 min, the demand is restored to 100% in another 3 sec. End-of-Life and Start-of-Life conditions are compared (*12*).

Thus the core power is reduced following the reduction in steam to the turbine, and the reactor is said to be load-following.

The transient calculation is performed in order to determine the system response (which depends on circuit delays and reactivity coefficients), core temperature overshoots (which depend on fuel time constants), and pressure changes in both the primary and the secondary circuits. Such a calculation will give design information to control engineers, fuel metallurgists, and pressure circuit designers, respectively.

As a later section will show, the LMFBR cannot load-follow in this manner because with sodium coolant and an added intermediate circuit the time delays are too long between the turbine and the core; also, there is no convenient negative coolant temperature coefficient to cause a power change of the correct sign.

1.5.2.2 *Accident Transients*

Figure 1.22 shows a reactivity addition to a PWR (*12*). The power rise resulting from the reactivity addition is curtailed by a reactor trip following a detection of excess mean channel temperature rise. This results in an immediate increase in the burn-out margin and safety for the PWR.

However, real protective systems cannot avoid delays and therefore, following the trip signal, a delay would result before the control rods are inserted. During this delay the burn-out margin would continue to decrease. The extrapolated curve shows that fuel burn-out would occur within 1.5 sec. This then is the allowable delay for this protective function. Such a calculation then has control implications and possible design implications for the control rods and the core.

A similar transient in the LMFBR would have very similar design motives. The only differences would be the result of smaller fuel time constants and the fact that failure would be marked by fuel melting and excessive cladding strain rather than burn-out (see Sections 2.2–2.4).

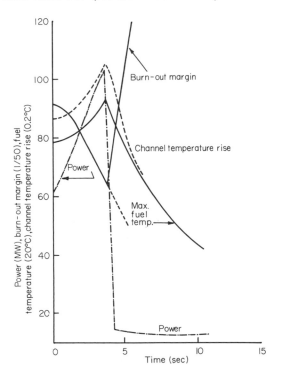

Fig. 1.22. A reactivity addition of 15¢/sec followed by a reactor trip of —$8 on excessive mean channel temperature rise (*12*).

1.5.3 STABILITY

The stability of a system can be assessed using a transfer function approach to give gain and phase margins to safety. These margins, being functions of system parameters, will have design implications. Section 2.5 deals with the assessment of stability in detail. Here we are concerned with instability time constants and with modes of instability as exhibited by a boiling water reactor (BWR).

1.5.3.1 *Time Constants*

Stability or instability is governed by a combination of time constants, such as circuit times or temperature responses of core components. Examples of the wide variety of effects that are possible are representative of the wide variety of time constants occurring in a reactor system.

(a) The reactivity feedback mechanism can produce instability directly if the overall feedback is positive. This is termed a "static" instability, occurring independently of time delays, although it is dependent on the time behavior of the temperatures providing feedback.

(b) Even with negative feedbacks, instabilities can occur. In a BWR, if the void transit time is comparable with fuel time constants, then the sequence of a power rise followed by temperature and void increase can multiply. It is an "oscillatory" instability with a time constant of a fraction of a second.

(c) Xenon instability occurs in very large thermal cores because a flux increase in one area is not quenched quickly enough by neutrons diffusing to other areas. Without selective area control (derived as the result of instability studies), a "spatial" instability involving oscillations of successive xenon poisoning and poison destruction cycles would result. This instability has a period of some 28 hr. It cannot occur in small fast cores.

1.5.3.2 *BWR instability modes*

A BWR is an interesting basis for any discussion on instability because so many different modes of instability have been identified. In practice, any instability would actually be a combination of two or more of the following:

a. *Simple voidage coupling.* This has already been referred to as oscillatory instability in Section 1.5.3.1. Reactivity, power, and voidage rise and, in combination with suitable delays, a reactivity decrease follows; if the delays are unfavorable, the reactivity may increase. A typical frequency would be in the range of 1 to 5 Hz.

b. *Hydraulic instability.*[†] This is nonnuclear and oscillatory with the following sequence of interactions. Following a disturbance to the inlet velocity (say an increase), the boiling boundary rises and the pump head thus required increases. Again, if delays in the circuits are unfavorable, then the resulting decrease in inlet velocity due to pumping inefficiency would come too late. Oscillations would occur with a frequency of about 1 Hz.

c. *Pressure variations.* A static instability, independent of time delays, was thought to exist. A pressure rise resulting in an inlet enthalpy decrease and thus a boiling boundary rise, would result in a reactivity, a power and thus a voidage increase, and then a pressure rise. The frequency would be of the order of 0.1 Hz. However, this effect is no longer important relative to the others.

d. *Ship's motion.* This would perhaps result in instability with a natural circulation BWR because the circulation head would be increased with an upward motion of the ship and vice versa. A 50% oscillation in apparent gravity could result in 100% oscillations in power and 10% in temperatures with a frequency of 0.1 Hz.

e. *Parallel channel instability.*[†] Such an instability would be exhibited between two or more channels having the same pressure difference across them and with the same common inlet and outlet headers so that flow can be divided between them in different fractions. The mass flow can oscillate between them, resulting in different two-phase pressure drops and voidages. It can only occur for exit qualities greater than 20–30%.

Sodium-cooled reactor systems are not subject to these types of instability and, indeed, a sodium-cooled system does appear to have excellent stability. However during accident conditions, unstable conditions might suddenly result. Section 4.4 will refer to sodium chugging after assembly failure, which is a form of static instability.

Stability studies are required to demonstrate that the particular system will be stable under normal and abnormal conditions. Such studies give information on such design features as the necessity for channel inlet gagging (in a BWR to make the total channel pressure drop less voidage dependent), pressure relief systems, or the avoidance of critical time constants in the circuit. Pump characteristics would also be tested for their satisfactory behavior following flow disturbances.

[†] See Davies and Potter (*13*).

1.6 Fault Tree Analysis

Fault tree analysis is a useful categorization tool with which the interrelationships between reactor components, their failures, and the reactor safety features can be defined. It provides a means for ensuring that safety analysis is all-inclusive, and it provides eventual potential for quantifying accident probabilities.

1.6.1 DEFINITIONS

A fault tree is a sequence of events which leads from one or more faults to the causes of those faults.

Systems analysts use such fault trees to: (a) define critical paths in the accident analysis; (b) calculate the probabilities of failures leading to given consequences *or* of consequences occurring in the system from one of a number of different initiating faults; and (c) specify safeguards against damaging consequences for each branch of the tree.

For different purposes the different trees available that will be discussed are: (a) a *single-failure tree* defined as a successive analysis of the causes of a single undesirable event; (b) a *multiple-failure tree* defined as an analysis of the consequences of a whole range of faults leading to a whole range of possible safe and unsafe terminations; and (c) an *accident-process tree* defined as a successive analysis of the consequences of a single fault. (This tree is a single branch of the multiple-failure tree and the reverse of the single-failure tree.)

1.6.2 SINGLE-FAILURE TREE

This is the most usually reported version of the fault trees.

1.6.2.1 *Logic Symbols*

The shorthand symbols which are used to define a fault tree are very few. They consist of a set of logic symbols which interconnect an undesirable event with its causes and at the same time define the relationship between the event and its causal events. As we shall see they are Boolean operators.

Figure 1.23 shows the three operators:

(a) AND gate: This describes the connection by which an event takes place only if *all* the input events occur.

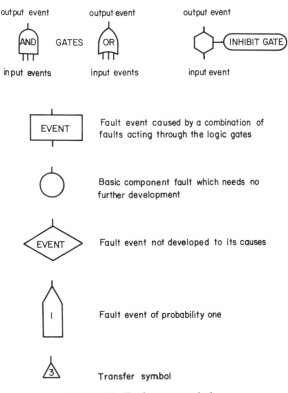

Fig. 1.23. Fault tree symbols.

(b) OR gate: This describes the connection by which an event takes place even if only one of the input events occurs.

(c) INHIBIT gate: This is a causal relationship which states that the output event is caused by the input event *if* the indicated condition is satisfied. It is a conditional connection.

In addition a set of symbols defines events in terms of their state of analysis within the fault tree. Figure 1.23 shows the fault event which is expanded from its causes to its consequences, the fault event which, as a basic component fault, requires no further analysis, and the fault event which, although it could be developed further, is not.

Two special symbols define a transfer of one part of the tree to another branch and an event which has unity probability of occurring (that is, a certainty).

These 8 symbols form the basic blocks from which a fault tree is constructed. All that remains is to connect the different events in a particular system.

1.6.2.2 *Shorted Motor Protection Circuit*

An electrical circuit is particularly simple to analyze by a fault tree because the connections are straightforward. Consider the protection circuit shown in Fig. 1.24 (*14*).

When the switch is closed, power is applied to the timer coil. This closes the timer contacts and applies power to the relay coil, which in turn closes the relay contacts. Power is then supplied through the fuse to the motor. When the switch is opened, the reverse procedure applies.

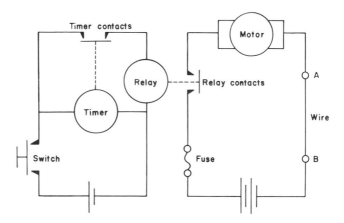

Fig. 1.24. A circuit diagram for a motor timing circuit in which a timer and fuse appear as safety devices (*14*).

The fuse and the timer are safeguards; if the motor fails shorted while the relay contacts are closed, then the fuse opens and shuts off the power, and if the switch fails to open again after some time (which is preset) then the timer will open its contacts and remove power from the motor.

The overheating of the wire A–B is an undesirable event in this circuit and it can be prevented if the safeguards operate. As an example of fault tree technique, the following paragraphs analyze the causes which might allow the overheating of A–B.

Figure 1.25 shows the basic fault tree for the undesirable event: *an overheated wire*. It could arise only as a result of excessive current *and* the power being applied for an extended time. The excessive current could come only if the motor fails shorted *and* the fuse fails to open, and the extended time during which this overpower is applied may result if either the relay contacts fail in a closed position, *or* the power is not removed from the relay coil.

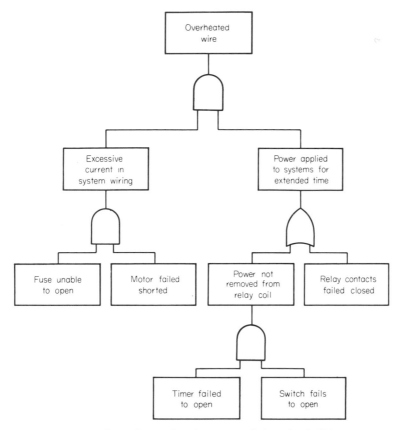

Fig. 1.25. Fault tree for the motor timing circuit (*14*).

The power is not removed from the relay coil only if both the timer fails *and* the switch fails to open. Thus the symbols demonstrate the logical connections between the events.

There are two analytical techniques for constructing fault trees. Each develops the tree to a different depth of detail. The first considers *primary component failures* that occur while the component is functioning within conditions for which it was designed. The second technique considers *secondary component failures* that occur when the component is subjected to abnormal conditions.

1.6.2.2.1 *Primary failure technique.* This method is used mainly in the communication field (control and instrumentation) and in data processing systems (control).

To illustrate the technique we may further analyze the causes of an overheated wire (Figure 1.26) (*14*). The *motor failed shorted* and *relay contacts failed closed* are both primary failures. To cause the *timer* to be *unable to open* there are two other primary causes: the *timer coil failed to open* and *the timer contacts failed closed*. If the *switch fails to open*, a primary failure of the *switch contacts failed closed* may be responsible, although there are other possible external causes which are here undeveloped.

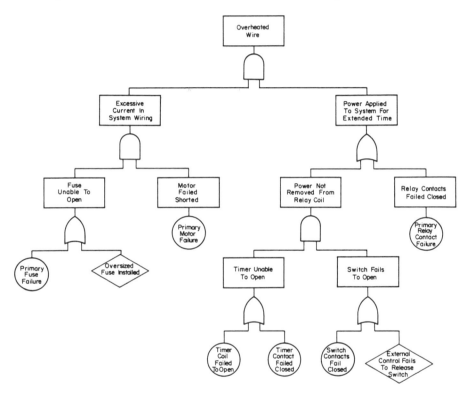

Fig. 1.26. Fault tree using primary failure technique (*14*).

1.6.2.2.2 *Secondary failure technique*. This method requires a greater insight into the system and its component parts because it needs a knowledge of how systems might operate and interact in abnormal conditions. It is used mainly in accident analysis and therefore it is of most concern to this discussion.

To illustrate this technique we can again further develop the example (Figs. 1.27a and 1.27b) (*14*). The *motor* and *relay* are sensitive to the failure of one another. In Fig. 1.26, *relay contacts failed closed* was a primary fault, but there may be a secondary reason; they may fail to open because

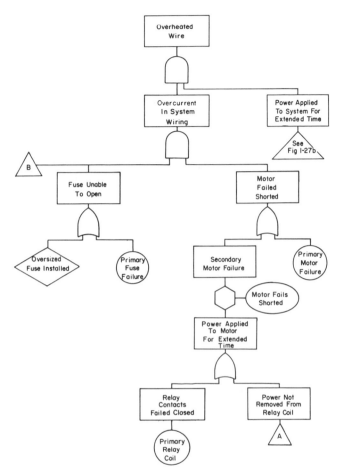

Fig. 1.27a. Fault tree using secondary failure technique (*14*). Branch in which there is an overcurrent in the system wiring, which helps to cause an overheated wire.

they have been fused shut by an overcurrent. The *overcurrent* is a cause of failure only *if* the *relay contacts fuse*. Therefore an inhibit conditional gate is needed to describe this situation. The overcurrent might arise if failure of the fuse occurred and the motor failed shorted.

The *motor failed shorted* might also come from a secondary (abnormal conditions) failure, an extended period of power, but *power applied for extended time* is a cause for failure only if the *motor* actually does fail shorted, so another inhibit gate is required. This extended power might arise from a failure of the relay contacts or power not being removed from the relay coil.

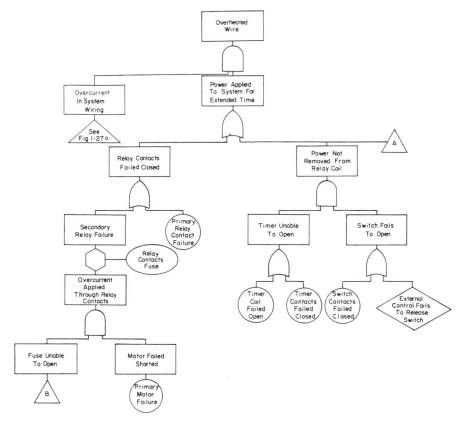

Fig. 1.27b. Fault tree using secondary failure technique (*14*). Branch in which the power is applied for an extended time, thus helping to cause an overheated wire.

Now we notice that the tree is no longer a simple structure but some faults occur in different parts of the tree; thus interconnections or transfers are required.

1.6.2.3 *Safety Features*

An AND gate is indicative of a safety feature, because two or more conditions must simultaneously be satisfied. In the example, these safety features are the timer and the fuse. Notice that the removal of the timer and the fuse, the safety features, would remove the AND gates.

AND gates also require a sequential operation of the events. The fuse must fail prior to the motor, and the timer must fail prior to the switch; otherwise, the consequences will be unrelated to the tree output.

An INHIBIT gate is similarly an indication of some safeguard, as there is only a conditional connection between the events. The undesirable event cannot occur, unless a certain condition is satisfied, as well as the input being present.

Thus a safe system will have a considerable number of AND gates and INHIBIT gates in its fault tree. A system with safety features in each branch would be demonstrably safe.

1.6.3 FLOW OR ACCIDENT-PROCESS TREE

This is another form of fault tree most commonly used in the analysis of the course of the accident. It proceeds in the direction opposite to the single-failure tree, from cause to a number of consequences rather from a consequence to its various causes.

Starting from the initiating fault, an attempt is made to see where the events may lead, just as the accident might progress. Therefore, consequences may be both safe and unsafe.

To illustrate a flow-process tree, consider Fig. 1.28. This follows the

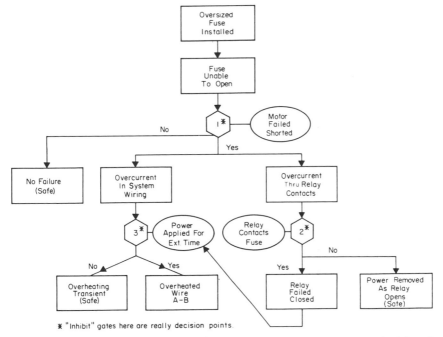

Fig. 1.28. An accident-process fault tree for the installation of an oversized fuse in the motor timing circuit.

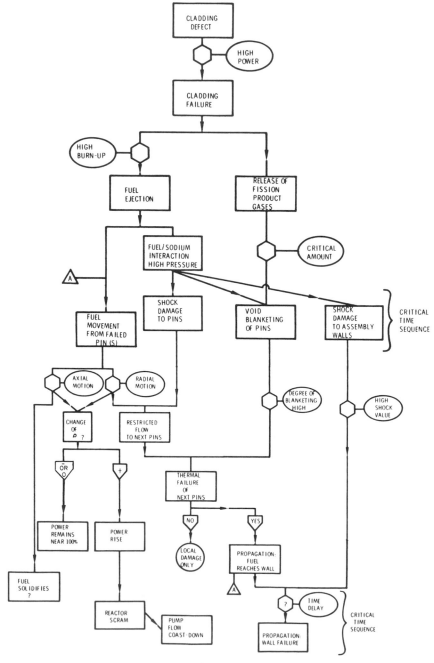

Fig. 1.29. An accident-process fault tree for a cladding defect and its consequences.

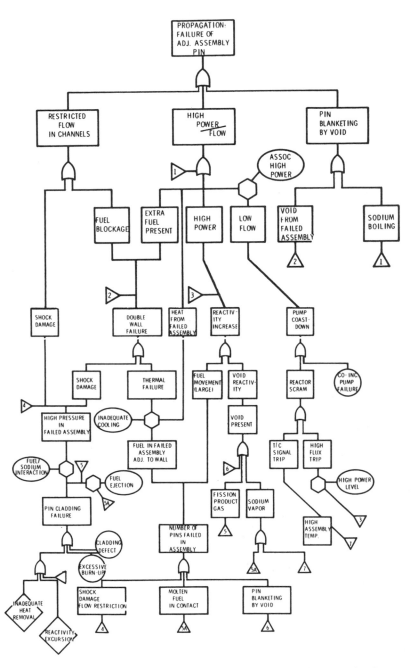

Fig. 1.30. A single-failure fault tree for the propagation of failure to pins in an assembly adjacent to one that has failed.

behavior of the particular system in the face of a reasonably probable fault: the installation of an oversized fuse. If an *oversized fuse* were *installed*, then the fuse would be unable to open in the face of excessive power. This is not a problem, unless the motor fails, although the failure is always a potential one while the oversized fuse is in position.

However if the motor does fail shorted, then an overcurrent will result through the system wiring and through the relay contacts. So the tree is further extended into two further possible terminations—one in which there is transient overheating but the wire is safe and the other in which the wire does fail.

Such a fault tree answers the question: What happens as a consequence of this event? It is the tree that one would follow in any calculation of the ensuing accident. There will be further examples throughout this book; a particularly good one lies in the analysis of fuel failure and its possible extension to further failures.

Figure 1.29 shows this tree in its accident process configuration while Fig. 1.30 shows the single-failure fault tree for the pin-failure propagation. They are equivalent, and they should be referred to again after reading Chapter 4. Because such trees require an intimate knowledge of the processes involved before they can be made unique, there is in fact great variety in how such trees may be laid out.

1.6.4 MULTIPLE-FAILURE FAULT TREE

Both of the preceding trees are really parts of a tree that has a multiplicity of roots and branches. This is called the multiple-failure tree. It starts with many faults and traces them through the various conditions and events to many terminations. It is more generally used as a survey of the whole of a given system that may then be broken down into critical and more detailed single-failure trees and accident-process trees. It is particularly useful in defining the interrelationship between events, systems, and safety features.

Figure 1.31 shows a multiple-failure tree devoted to four major faults that might occur in a sodium-cooled system. Terminations are shown to be mainly of the safe variety (including transient but undamaging overheating), although a major accident to the core is another termination shown. Safety features abound to inhibit the accident from reaching this termination. Also shown is the importance of the plant protective system in providing for the detection of the abnormality and rapid shut-down. It is the one safety feature that applies across the board to all the faults considered.

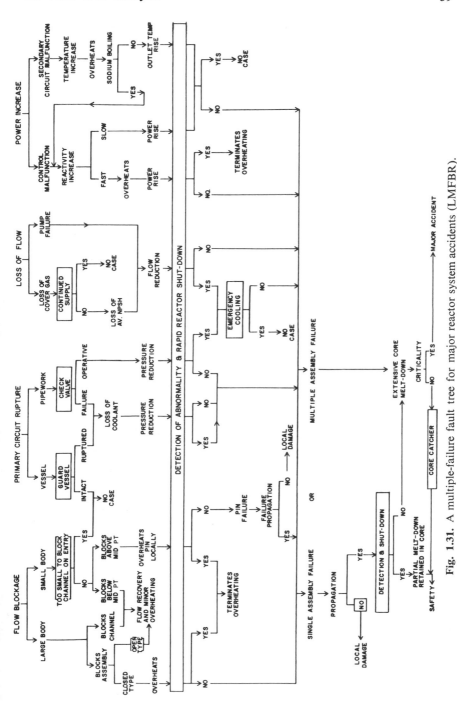

Fig. 1.31. A multiple-failure fault tree for major reactor system accidents (LMFBR).

1.6.5 FAULT TREE PROBABILITY EVALUATION

A quantitative evaluation of a fault tree will establish the likelihood of the undesired event and will say what the relative contributions are to this likelihood from each failure mode. By knowing such a likelihood, it would be possible to decide whether or not corrective action is needed or warranted.

In fact, an engineer can make an evaluation of a fault tree, using his experience to discriminate among more and less likely failure modes, and he does indeed decide whether to take corrective action by assessing relative probabilities according to his engineering judgment.

If, however, the probabilities of the different initiating faults are known, it is possible to make a strict numerical fault tree quantization. There are two methods of evaluation, computation and statistics, and each method has two further submethods (*15*).

(1) Computation:
    (a) direct fault tree calculation,
    (b) algebraic expression evaluation.

(2) Statistics:
    (a) Monte Carlo simulation,
    (b) importance sampling simulation.

One can illustrate the use of the two major methods by pointing out that the probability of turning up a 7 from a throw of two dice can be obtained by calculating it or by throwing a pair of dice a large number of times and counting. Here we are interested in the computational methods.

1.6.5.1 *Boolean Quantization*

By defining an algebra to quantize each logic gate, we can calculate a probability for the final consequence in a single-failure fault tree from the initial probabilities of each causal event.

The AND gate is defined as a multiplicative operator; with inputs X_1, $X_2, \ldots, X_n$, the output of the AND gate is the product $X_1 X_2 \cdots X_n$.

The OR gate however is defined as an additive operator; with inputs $X_1, X_2, \ldots, X_n$, the output of the OR gate is the sum $X_1 + X_2 + \cdots + X_n$.

The INHIBIT gate is also defined multiplicatively; if an input X_n has a condition W then the output of the INHIBIT gate is the product WX_n.

Figure 1.32 shows the Boolean expression for each gate. The INHIBIT and AND gates are clearly very closely related. In many cases the fault tree

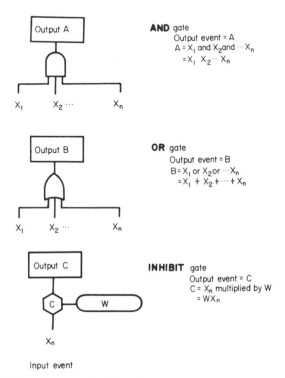

Fig. 1.32. Boolean quantization of AND, OR, and INHIBIT gates.

can be expanded by the use of either interchangeably; it depends on whether the analyst wishes to probe the reasons for the INHIBIT conditional modifier.

Figure 1.33 shows a representative section of a fault tree with inputs X_1, X_2, and X_3. The final output event may be quantized by evaluating each gate output thus:

$$A_1 = B_1 A_2 \qquad B_2 = X_1 + X_2 \qquad B_3 = X_1 + X_2$$
$$B_1 = X_2 + B_2 \qquad A_2 = B_3 X_3 \qquad (1.60)$$

Then the overall output is given by

$$\text{Output} = A_1 = B_1 A_2 = (X_2 + B_2)(B_3 X_3) = (X_2 + X_1 + X_2)(X_3)(X_1 + X_2)$$
$$= X_1 X_2 X_3 + X_1 X_1 X_3 + X_1 X_2 X_3 + X_2 X_2 X_3 + X_1 X_2 X_3 + X_2 X_2 X_3$$
$$(1.61)$$

$$= 3 X_1 X_2 X_3 + X_1 X_1 X_3 + 2 X_2 X_2 X_3 + \text{?}$$

There are two types of redundancies that can be used to simplify this

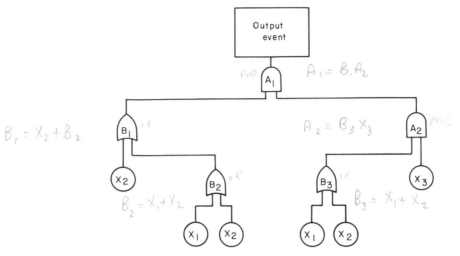

Fig. 1.33. Boolean quantization of a typical fault tree.

equation. The redundancies apply because Boolean combinatorial logic is binary, variables in the logic can only be one or zero.
Thus the AND redundancy states that

$$A \cdot A = A \tag{1.62}$$

The OR redundancy states that

$$B + B = B \tag{1.63}$$

and so also

$$B + B \cdot C = B \tag{1.64}$$

The logic of each redundancy can be checked by considering the original definition of the two gates. For example, if an output only applies when both of two identical inputs apply, then it is the same as if there were only one input.
Applying the AND redundancy, Eq. (1.61) becomes

$$\text{Output} = X_1 X_2 X_3 + X_1 X_3 + X_1 X_2 X_3 + X_2 X_3 + X_1 X_2 X_3 + X_2 X_3$$

and then, removing the OR redundancies, the expression is reduced to:

$$\text{Output} = X_1 X_3 + X_2 X_3 \tag{1.65}$$

This is now a nonredundant Boolean expression of the combinatorial logic of the fault tree in Fig. 1.33. It can be used to derive a probability of the output event from the probabilities of the individual input events X_1, X_2, and X_3.

1.6.5.1.1 *Probability calculus.*[†] The Boolean expression is transformed into a true probability expression using the following relationships:

(a) $P(AB) = P(A)P(B)$ (1.66)

The probability of a number of events all occurring is a multiplication of the separate probabilities of each event.

(b) $P(A + B) = P(A) + P(B) - P(A)P(B)$ (1.67)

The probability of one of several events occurring is the sum of the separate probabilities less the probability of their occurring together. If the probabilities of each event are small then the last term in this expression can be omitted.

Similar expressions apply for three or more variables.

Thus the probability of the final output in the fault tree of Fig. 1.33 is given by a transformation of the logic expression of Eq. (1.65) into a probability expression:

$$P(\text{Output}) = P(X_1)P(X_3) + P(X_2)P(X_3) - P(X_1)P(X_3)P(X_2)P(X_3) \quad (1.68)$$

1.6.5.1.2 *Repair time.* It is not enough to know the probability of each event, as each failure may remain in operation for a different repair time. Thus an infrequent failure rate of a system may be associated with a long repair time, and this may be more significant than a high failure rate of a system associated with a very short repair time.

It is conservative to assume that nothing is repaired. It is indeed very pessimistic and unrealistic for a minor event that has a high probability of occurring. No really satisfactory solutions to the problem of mathematically representing the repair time have been found, although computer programs that assume constant failure rates and constant repair times do achieve very close approximation to the true results for small fault trees (200 inputs or less). Such programs are used in the aircraft industry where the probability of initiating faults in protection systems is well known and can be applied to reactor systems to advantage (*16*).

[†] See Roberts (*15*).

A constant failure rate assumes that the system is operating in the flat portion of the life reliability "bathtub" curve when the burn-in period has ended and before the wearout period becomes apparent (Fig. 1.34). This is a reasonably good assumption for systems that are tested throughout the burn-in period and that are replaced before the wearout period commences; in particular, it is reasonable for certain electrical control equipment.

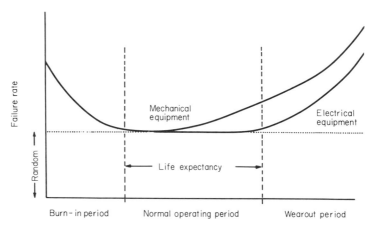

Fig. 1.34. Life characteristic curve for mechanical and electrical equipment.

1.6.5.2 *Simulation*

The other method of obtaining the probability of a fault tree output event is by statistical methods, which include simulation. The various logic gates are simulated on a computer, even to incorporing a repair time generator with a randomly generated fault sequence. The simulator then traces the mock use of the system for a large number of computer generated faults.

One severe disadvantage of this method of evaluation is the large number of trials required. The probabilities to be demonstrated in the fault tree may range to 10^{-10} and below. For verification of a probability of 10^{-6} at 10% confidence 105,000 trials are required if no output fault occurs in the simulated trials. For 90% confidence 2,300,000 trials would be needed. However if one fault occurs in the simulated trials, 530,000 and 3,900,000 trials would be needed, respectively, to obtain 10 and 90% confidence in the 10^{-6} result produced.

Then, for a computer taking 10 μsec for each increment of the fault tree, where the fault tree requires 10^4 increments per trial, a million trials would take 28 hr of computer time. Such a calculation is prohibitively expensive. For this reason, new Monte Carlo methods are being produced to obtain simulated statistical probabilities.

1.6.5.3 *Probability in Fast Reactor Safety*

Probability analysis is best used where failure statistics are good, for example, in control systems that are made up largely of conventional components. To date, probability analysis is not used regularly in accident analysis.

However, pseudo-probability analysis, otherwise known as engineering judgment, has been supplemented in some cases by statistical analysis and by partial probability analysis. For example, failure of scram systems has been variously estimated at between 10^{-4} and 10^{-6} per reactor year despite the difficulty of quantizing possible short-circuit paths. On a somewhat firmer basis (*17*), from an analysis of safety and shim rod hang-ups in the SM-1A reactor at Fort Greeley in Alaska during a 72-month period, the probability of failure to scram on a two out of three system is calculated to be $4.1 \cdot 10^{-4}$ per reactor year. Such an analysis applies to a particular reactor system but in general can be applied to any plant protective system composed of similar parts.

On a somewhat more speculative basis and applying now directly to the sodium-cooled fast reactor, one analyst has applied the figure of 10^{-8} per reactor year to the probability that a small local failure in a fuel subassembly would propagate to a neighboring assembly. The error on this probability figure would be as much as two orders of magnitude either way!

In general then, the fast reactor is in the same position as the thermal systems: there is a desire to use probability theory to aid in the assessment of reactor safety, and a good deal of work is being done to apply the theory to protective systems and systems that use components for which failure rates are reasonably well known (*18–20*). In addition, further effort is being applied to overall containment analysis, where quantitative engineering judgment is used in place of failure rate data (*21, 22*).

When failure rates for reactor components become known (see Section 3.1), then probability analysis of system safety will undoubtedly be an established assessment method. Such a method will clearly be used initially to improve the reliability of reactor subsystems, such as the protective system or emergency core-cooling system (*19*), but ultimately general safety accident analysis should benefit by an application of these methods.

REFERENCES

1. P. J. Searby, The growth of demand for energy and the role of nuclear power. *IAEA Symp. Nucl. Energy Costs Econ. Develop.*, Istanbul, 1969 (reprinted in *Atom*, November 1969).

2. Fuels—A special report. *Power* **112** (6), 196 (1968).

3. S. Glasstone and M. C. Edlund, "The Elements of Nuclear Reactor Theory." Van Nostrand, Princeton, New Jersey, 1966.

4. G. R. Keepin, T. F. Wimett, and R. K. Zeigler, Delayed neutrons from fissionable isotopes of uranium, plutonium and thorium. *Phys. Rev.* **107** (4), 1044–1049 (1957).

5. R. Gwin, Determination of reactor parameters from period measurements. Annu. Rept. period ending September 10, 1956, ORNL 2081, p. 84. Appl. Nucl. Phys. Div., Oak Ridge Nat. Lab., Oak Ridge, Tennessee.

6. J. K. Long, W. B. Loewenstein, C. E. Branyan, G. S. Brunson, F. S. Kirn, D. Okrent, R. E. Rice, and F. W. Thalgott, Fast neutron power reactor studies with ZPR III. In *Proc. 2nd United Nations Conf. on Peaceful Uses At. Energy, Geneva, 1958. A/Conf. 15/Paper 598.*

7a. J. H. Perry, ed., "Chemical Engineers' Handbook," 3rd ed. McGraw-Hill, New York, 1950.

7b. C. B. Jackson, "Liquid Metals Handbook," 3rd ed. U. S. Govt. Printing Office, Washington, D. C., 1955.

8a. "Modern Computing Methods." Notes on Appl. Sci., No. 16, Dept. of Sci. and Ind. Res., Nat. Phys. Lab. H. M. Stationery Office, London, 1957.

8b. L. S. Tong, "Boiling Heat Transfer and Two-Phase Flow." Wiley, New York, 1967.

9. R. W. Bowring, Physical model based on bubble detachment, and calculation of steam voidage in the subcooled region of a heated channel. OECD Halden Reactor Project Report HPR 10, OECD, December 1962.

10. R. W. Tilbrook and G. Macrae, PASET, A transient code for plant analysis of sodium voiding. *Trans. Amer. Nucl. Soc.* **12**, 904 (1969).

11. M. J. Driscoll, Notes on fast reactor physics. *Nucl. Power Reactor Safety Course, M.I.T., Cambridge, Massachusetts, Summer 1969.*

12. J. Graham, Notes on reactor kinetics and stability. RS/L 110 (Revised). UKAEA Postgraduate Educ. Centre, Harwell, England, May 1967.

13. A. L. Davies and R. Potter, Hydraulic stability—An analysis of the course of the unstable flow in parallel channels. Brit. Rept. AEEW R.446. U. K. Atomic Energy Authority. Dorchester, 1966.

14. D. Haasl, Advanced concepts in fault tree analysis. *System Safety Symp. Seattle, Washington, June 1965.*

15. N. H. Roberts, "Mathematical Methods in Reliability Engineering." McGraw-Hill, New York, 1964.

16. M. E. Stewart, J. F. White, and J. O. Zane, Reliability analysis of the power burst facility reactor protection system. USAEC Rept. IN-ITR-200. U.S. At. Energy Comm., Idaho Falls, Idaho, December 1969.

17. D. Fitzgerald, C. Feavyear, and G. Knighton, A probabilistic evaluation of a nuclear plant protection system. *Amer. Nucl. Soc.* **12**, Supplm. p. 2, 1969.

18. B. J. Garrick and W. C. Gekler, Reliability analysis of engineered safeguards. *Nucl. Safety* **8** (5), 470 (1967).

19. I. M. Jacobs, Reliability of engineered safety features as a function of testing frequency. *Nucl. Safety* **9** (4), 303 (1968).

20. "Failure Rate Data Book," Vol. 3, Tri-Service and NASA failure rate data (FARADA) Program. Naval Fleet Missile Systems Analysis and Evaluation Group, Corona, California, September 1968.

21. F. R. Farmer and E. V. Gilby, A method of assessing fast reactor safety. *Int. Conf. Safety of Fast Breeder Reactors, Aix-en-Provence, September 1967*, Paper VI-2-1.

22. L. Cave and J. R. Crickmer, The safety assessment of fast reactors by probability methods. *Int. Conf. Safety of Fast Breeder Reactors, Aix-en-Provence, September 1967*, Paper VI-3-1.

GENERAL REFERENCES

J. G. Yevick and A. Amorosi, eds., "Fast Reactor Technology: Plant Design." M.I.T. Press, Cambridge, Massachusetts, 1966.

T. J. Thompson and J. G. Beckerley, eds., "The Technology of Nuclear Reactor Safety," Vol. I, "Reactor Physics and Control," M.I.T. Press, Cambridge, Massachusetts, 1964 and Vol. II, "Reactor Materials and Engineering," M.I.T. Press, Cambridge, Massachusetts, 1971.

W. B. Cottrell, Nuclear Safety, A Bimonthly Tech. Progr. Rev., Oak Ridge National Laboratory, Oak Ridge, Tennessee.

CHAPTER 2

SYSTEM DISTURBANCES

2.1 Description of Main Reactor Systems

In order to discuss possible system disturbances comprehensively, it is convenient to relate these disturbances to three distinct fast reactor systems. These three systems are representative of fast reactor types and moreover cover a wide range of coolant pressures and fuel and coolant types, thus making the discussion more general. Although the present book is intended to particularize toward the sodium-cooled breeder, it is nevertheless important to know what the alternatives are and how they behave. The major alternatives are the gas-cooled and steam-cooled systems.

2.1.1 LIQUID-METAL-COOLED FAST BREEDER REACTOR

The favored fast reactor system in the United States, Great Britain, the USSR, and France—this reactor is now cooled by sodium although NaK systems have been used. Designs generally use mixed oxide fuels in a pin configuration cooled by upflowing coolant to remove anything between 12 to 16 kW/ft (1).

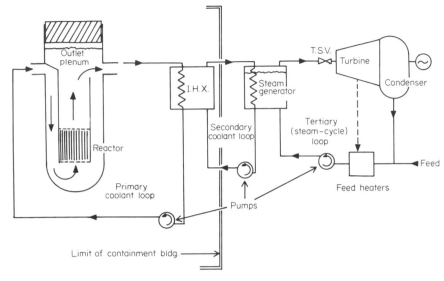

Fig. 2.1. An indirect LMFBR heat transport system with an intermediate loop showing the limit of the containment enclosure.

TABLE 2.1

FAST REACTOR TYPES

Reactor characteristic	Liquid-metal-cooled breeder	Gas-cooled breeder	Steam-cooled breeder
Fuel type	(Pu, U) oxide	(Pu, U) carbide	UO_2
configuration	0.25 in. pin OD	1 mm particles	cermet or pins
cladding	SS	silicon carbide	SS
rating/density	12–16 kW/ft	240 W/cm³	—
fission product retention	vented or unvented	retained	retained
Core L/D ratio	0.33–0.5	0.75	0.6
Coolant	sodium	helium	steam
phase	liquid	gas	supercritical
inlet temperature	750°F	570°F	750°F
outlet temperature	1000°F	1830°F	1050°F
pressure	15–50 psia	750 psia	3700 psia
Heat transfer system			
configuration	loop or pool	loop	loop
cycle	3	direct?	direct
Reference	*(1)*	*(2)*	*(3)*

Temperatures in the system range from 750°F at inlet to 1000°F at the mixed core outlet, while the coolant is kept at a low pressure of 1 atm or a little above.

The heat transfer systems are usually all three cycle systems with an intermediate or secondary system to insulate the radioactive primary sodium from the turbogenerator steam cycle. However, present designers are divided on whether the reactor should be a pool or a loop system. The former immerses the core, heat exchanger, and primary pumps in a single pool of sodium, while the latter uses pipes to connect its primary components and its secondary components. There are advantages and disadvantages for both systems (see Section 4.1) but for the present discussion a loop system is assumed (Fig. 2.1). Table 2.1 shows the main characteristics of the liquid-metal fast breeder reactor (LMFBR).

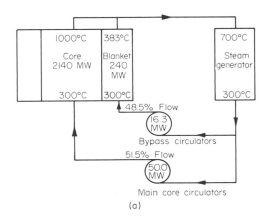

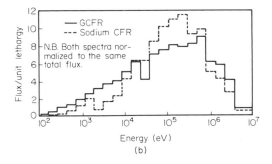

Fig. 2.2. Gas-cooled fast reactor conditions: (a) coolant flow and temperature conditions; (b) comparison of GCFR neutron spectrum with that of an LMFBR (*2*).

2.1.2 Gas-Cooled Fast Reactor

This system is the second runner in both the United States and Europe, with possibly greater economic potential than the liquid-metal-cooled variety. However, it still has development problems to overcome (Fig. 2.2).

The particular version discussed here is a British design (2) which uses coated particle technology to produce a novel fuel that allows very high outlet temperatures in the range of 1000°C (1830°F). The fuel uses small silicon carbide coated particles of mixed carbide fuel which retain the developed fission³ products. The power density is in the range of 240 W/cm³. The coolant is helium operating at a pressure of above 750 psia, and potentially the system has promise in direct cycle use with the addition of a gas turbine. Table 2.1 shows the main chacteristics for a 1000 MWe version of this reactor system.

2.1.3 Steam-Cooled Fast Reactor

Work on this system has now largely ceased in the United States and in Europe. However, it does have very different characteristics from either of the other systems and is worth including (Fig. 2.3).

The design considered could employ either stainless steel clad pins or stainless steel cermet. The coolant is supercritical steam; therefore the

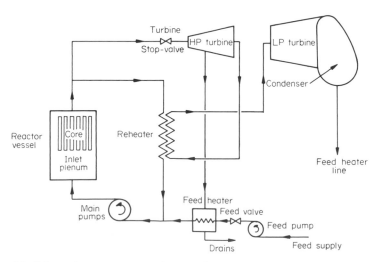

Fig. 2.3. Schematic arrangement of steam flow in a postulated supercritical steam-cooled fast reactor (3).

system pressure is very high in the range of 3700 psia. Temperatures are very close to those of the LMFBR system.

With steam coolant a direct cycle system is the natural design configuration, although steam cycle efficiencies may demand one or more stages of reheat. Table 2.1 shows the main characteristics of the supercritical steam-cooled system.

2.2 Flow Perturbations

It is clear that these three systems will have very different responses to flow perturbations and because of the high pressures involved, flow perturbations in the gas- and steam-cooled systems due to system rupture are classified as depressurizations (Section 2.2.4).

2.2.1 SYSTEM MODELING

The modeling of the coolant flow has been treated in the first chapter. The flow enters the core at a low temperature and is raised in temperature through the core. It then transfers the removed heat to a heat sink in the form of a heat exchanger or a turbine.

While in the core, the heat balance in the coolant is represented by an equation of the form

$$m_c c_c \, \partial T_c / \partial t = h_f (T_f - T_c) - m_c c_c v_c \, \partial T_c / \partial z \tag{2.1}$$

where the heat-transfer coefficients are flow dependent, according to some correlation between the Nusselt and the Reynolds and Prandtl numbers:

$$N_u = a + 0.025 (R_e)^{0.8} (P_r)^{0.8} = h_f d_e / kA \tag{2.2}$$

Thus a reduction of flow reduces the heat removal term in Eq. (2.1) directly through the velocity v_c and so T_c increases. The flow reduction also decreases the heat transfer h_f and the fuel temperature T_f increases by more than just the increase in T_c. It is noted that the coolant temperature increases first and is a primary indication of flow reduction.

Flow perturbations arise from: (a) system malfunctions such as pump failures, a loss of system pressure, or blockages which might arise from isolation valve malfunction; and (b) local blockages within the subassembly or at its inlet. The latter cases of local blockages are treated in Section 4.4; this section is concerned with overall system malfunctions.

2.2.2 Pump Failures

A flow disturbance can result from either a loss of power to the pump motors causing the pump to coast down or a mechanical failure of the pump. The former coast-downs can occur in one or more pumps simultaneously depending only on how the pump motor electrical supply lines are interconnected. However, a mechanical failure of a pump is expected to occur in only one pump at any time since it is a very unlikely fault.

The loss of electrical supply may occur in two or three primary pumps or in one primary and a secondary pump, just what happens to the core flow depends on a detailed flow balance in the system. Flow conservation equations are used together with assumed pump characteristics to derive core flow as a function of time t. Such flow coast-down curves take the form:

$$\text{relative flow} = 1 - [t/(t + t_1)] \tag{2.3}$$

In a typical pump failure in an LMFBR due to a loss of electrical power the time constant t_1 is of the order of 2 sec, while a pump seizure would be somewhat faster with a t_1 of the order of 1 sec.

The core flow due to a mechanical failure of a pump in one loop of a two-loop plant might typically take the form:

$$\text{relative flow} = a + b \tan(c - dt) \tag{2.4}$$

where a, b, c, and d are constants.

This core flow behavior is input to a calculational model to predict changes in coolant and fuel temperatures with and without a reactor trip in the event of either of these flow perturbations. Figure 2.4a shows a typical flow rundown due to a loss of all electrical power to the pumps of a LMFBR, while Fig. 2.4b shows the resulting core temperatures.

It will be noted in Fig. 2.4b that the initial temperature rise is cut back as the reactor is scrammed, but it rises again as the flow drops rapidly to 5% or less than the decay power level. However the temperatures again decrease as the power decays still further and becomes less than the flow level relative value.

Notice that the monitored temperatures are the coolant and cladding temperatures as the former first changes due to the flow decrease and the failure of fuel pins would be the result of excessive cladding temperatures.

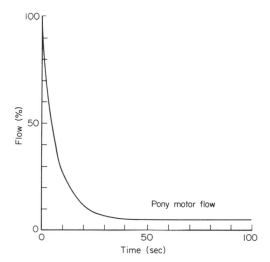

Fig. 2.4a. Primary LMFBR coolant flow resulting from a loss of electrical power to all primary pumps. Flow reduces to pony motor flow.

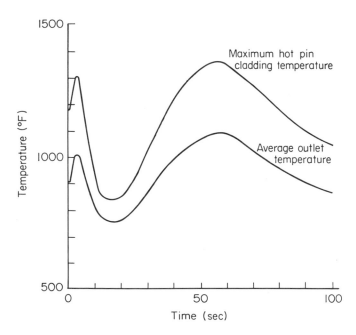

Fig. 2.4b. Reactor temperatures arising from a reduction of flow to pony motor flow resulting from a loss of electrical power to all primary pumps.

2.2.2.1 *Trips*

In the event of a flow perturbation the following trip signals would be available: (a) primary cause signals, electrical supply board power loss, or pump speed indication; (b) flow (say 85%) and power-to-flow ratio; and (c) outlet temperature. These trips are in order of occurrence. Figure 2.4b assumes a flow trip.

2.2.3 Loss of Flow Due to Pipe Rupture

In a LMFBR the effect of a loss of primary system integrity is a loss of coolant and a loss of core flow, whereas in the gas-cooled and the steam-cooled systems the effect can better be represented as a depressurization.

The calculation of core flow in a LMFBR after a pipe rupture is again a hydraulic balance. The effects are very similar to a flow coast-down but more severe. Figure 2.5 shows a comparison of the flow reduction due to a coast-down and a pipe rupture. The problems are the same; the time scales are different.

However, because in a pipe rupture the system is also losing coolant, it is necessary to make sure that there is sufficient sodium to maintain a cooling circuit. In a loop system this means that the sodium must not drain down to a level such as to cut the main cooling circuit, but in a pool system, despite a primary line break (say between the intermediate heat exchanger and the core inlet), adequate sodium is always provided although the flow rate is reduced.

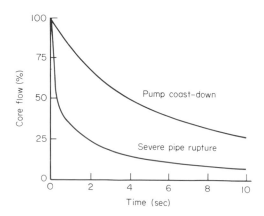

Fig. 2.5. Typical core flow rate following a severe pipe rupture in a loop-type LMFBR compared to a flow coast-down after pump trip.

2.2.4 DEPRESSURIZATION EFFECTS

These effects are very different in each of the three reactor types.

2.2.4.1 *LMFBR*

In the sodium-cooled system the main effect of depressurization (which may arise from a pipe rupture or a loss of cover gas pressure) is to reduce the pump suction head.

The normal pump characteristic curve (Fig. 2.6) shows the relationship between the mass flow M and the pump head as a function of pump speed w:

$$\text{pump head} = aw^2 + bwM + cM^2 \tag{2.5}$$

The figure also shows the system resistance and the normal operating point for the primary system. At this operating point the available net positive

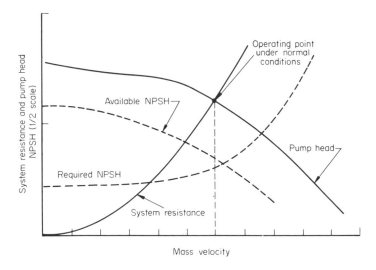

Fig. 2.6. The pump characteristics and required net positive suction head compared to system characteristics for an operating pump (*5*).

suction head (ANPSH) must be larger than the required net positive suction head (RNPSH) for that pump.

The behavior is different in the case when system pressure has been reduced, for the pump may now be cavitating. The pump characteristic now varies considerably (*5*).

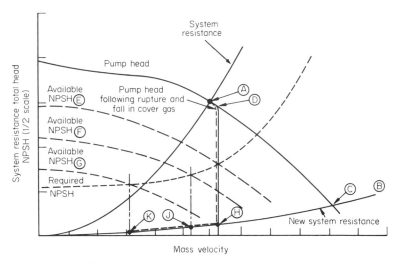

Fig. 2.7. Change of pump characteristics and operating point due to cavitation following a severe pipe rupture in a loop-type LMFBR (5).

Figure 2.7 shows predicted behavior following a large pipe rupture. The pipe rupture is first seen as a loss of system resistance and this curve rapidly falls to a new lower position B. The pump attempts to move to a new operating position C by increasing flow and moving down its characteristic

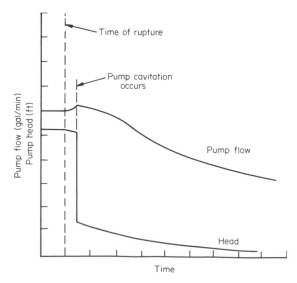

Fig. 2.8. Pump flow and pump head following a severe pipe rupture in a loop-type LMFBR (5).

at constant pump speed. However as soon as it moves to that flow where the ANPSH falls below the RNPSH, then the pump cavitates at D. Meanwhile, due to the loss of system pressure, the ANPSH curve has been decreasing from E to F to G. In order for the pump to operate to balance the system resistance and at the same time always maintain a RNPSH less than or equal to the available head, the behavior of the pump follows the curve A to D to H to J to K. The final flow is very low, satisfying the new system pressure conditions. Figure 2.8 shows transient conditions.

This effect and its representation complicates the prediction of core flows in the analysis of a pipe rupture in this fast reactor system (Section 2.2.3).

2.2.4.2 *Gas-Cooled Fast Reactor*

As a result of a system depressurization many of the parameters of Eq. (2.1) are changed because of the sensitivity of the coolant properties to the pressure. Thus m_c, c_c, and h_f (P_r and ϱ) all change.

A loss of pressure is exhibited principally in changes in density, and as m_c decreases, the heat input per gram of coolant is increased, thus increasing the temperature T_c if the power is maintained constant. The loss of system pressure has simply made the coolant less effective in removing the necessary heat.

Figures 2.9 and 2.10 illustrate the behavior of the gas-cooled reactor for a slow depressurization (2) under the following assumptions: (a) failure of an 8-in. diameter standpipe; (b) pressure decreases with a 78-sec time

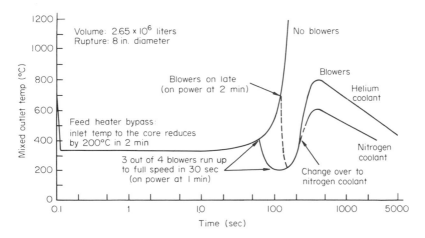

Fig. 2.9. Variation of mixed outlet temperature from a gas-cooled fast reactor following a depressurization accident (2).

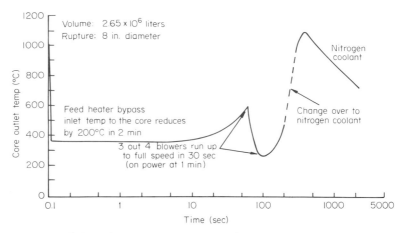

Fig. 2.10. Variation of core outlet temperature from a gas-cooled fast reactor following a depressurization accident (2).

constant; (c) reactor trip operates from a loss of pressure signal; and (d) power to the circulators fails and they run down to half speed in 15 sec and to 20% speed in 60 sec. The result of this transient would be quite unacceptable and therefore the following safeguards are assumed to apply in this case:

(a) Over the first two minutes after the trip, the boiler feed heating is reduced, and the inlet temperature drops from 300°C to 100°C.

(b) After one minute, emergency power supplies manage to restart 3 out of 4 circulators.

(c) Rupture disks fill the vessel with nitrogen when the system pressure has decreased to 2 atm; thus the coolant is changed to another which is 40% more effective in heat removal.

With these safeguards the temperature transient is held at acceptable values until the decay heat has reached a low enough value. These safeguard features are not unique but are representative of those which might be used in such an incident. Others have been suggested (4).

2.2.4.3 *Steam-Cooled Fast Reactor*

With a system pressure of up to 3700 psia, depressurization in this system can be a very severe accident, and it therefore requires immediate limitation by ensuring coolant boundary integrity and by designing a system in which only limited apertures have even the smallest probability of occurring.

The main effect of depressurization, apart from mechanical damage, is to reduce the coolant density. Depressurization rates of up to 200 psia/sec would perhaps be involved, with corresponding rates of change in coolant temperatures.

Another effect of the sensitivity of high pressure gas coolants to the level of pressure is that, even during the more probable flow coast-down incidents, the accompanying pressure changes can complicate the system behavior. Thus coolant temperatures rise both because of a loss of coolant effectiveness and because of a loss of coolant flow. The actual response depends critically on the density coefficient and its sign, but in general, system response is more rapid than in the sodium-cooled systems.

2.2.5 SECONDARY AND TERTIARY FLOW FAILURES

Such flow failures also disturb the primary system and the flows in the secondary and tertiary circuits would be calculated from hydraulic balances and pump characteristics in the same way as previously discussed. However, the disturbances are exhibited in the primary system as thermal changes due to a decrease in heat removal capability. Thus the core first experiences a rise in inlet temperature. As thermal disturbances, these effects will be discussed in Section 2.4.

2.3 Reactivity Perturbations

Reactivity perturbations have very much the same effect in any of the three reference systems except that the feedback coefficients may differ in value and sign. In all cases the reactivity disturbance is first noted as a rise in power and fuel temperatures and later by a rise in coolant temperatures. The Doppler coefficient is therefore most important in curtailing the transient.

2.3.1 SYSTEM MODELING

Section 1.4 showed that reactivity changes arise through perturbations to Σ_{fission}, Σ_a, and through changes in leakage. The reactivity variations are input into the calculational model through k_{eff} in the kinetic equations.

Such cross-section and leakage changes can arise from: (a) feedback effects from temperature changes, pressure changes, and structural movement; and (b) external influences from control rod movement, the addition of poisons or moderator due to component failure, or core voiding in the sodium system. The magnitude of the reactivity effects from each cause is the concern of the physicist who evaluates them, using steady-state codes, as if the changes were a set of pseudo-steady states. These reactivity changes are then used on the presumption that the changes in reactivity occur more rapidly than the initiating mechanisms.

Thus the reactivity input is:

$$\delta k(t) = f(t) + \alpha_i \, \delta T_i(t) \tag{2.6}$$

where $f(t)$ represents external influences and $\delta T_i(t)$ symbolizes temperature and pressure changes which give rise to feedback effects. The latter are linked to relevant model equations which produce those temperatures and pressures.

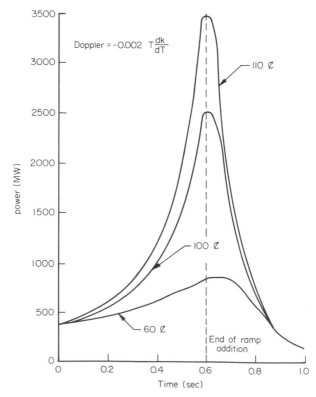

Fig. 2.11. The effect of various reactivity additions to a LMFBR operating at full power. The reactivity is added as a ramp terminated in 0.6 sec. Power variations are shown as a function of time.

2.3.2 CONTROL ROD MALFUNCTIONS

The reactivity inputs due to control rod malfunctions are limited both in the rate of addition and in the magnitude of the addition due to the design of the control rods themselves. It is usual to consider control rod malfunctions that add reactivity while the reactor is at power and the system is hot, and also during the start-up procedure.

2.3.2.1 *At Power*

When the system is hot, thermal feedbacks are immediately available, because any power change will produce a significant fuel temperature change.

The following trip signals would be available: (a) control rod drive sensors; (b) period meters if they are included in the system; (c) high flux; and

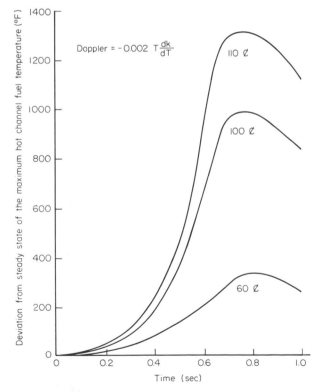

Fig. 2.12. The effect of various reactivity additions to a LMFBR operating at full power. The reactivity is added as a ramp terminated in 0.6 sec. Fuel temperature variations are shown as a function of time.

(d) high coolant outlet temperature eventually. Knowing the trip signals available and the delays between the monitored parameter reaching the trip value and the rods commencing to move into the core, it is possible to define a highest safe rate of addition and a largest acceptable step addition of reactivity. In a typical LMFBR a step addition of 60 cents may be accommodated and rates of up to $ 3 or $ 4/sec could be acceptable. Naturally it is also possible to design the control system to do better than this if needed by shortening the delays in the electronics and accelerating the rods when inserting the shut-down absorbers.

Figures 2.11 and 2.12 show the effect of adding terminated ramps of reactivity to a typical LMFBR in terms of its power rise and increase in

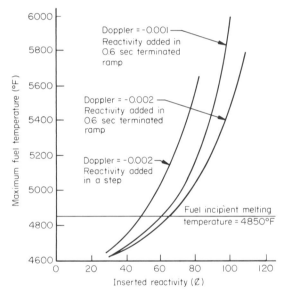

Fig. 2.13. The maximum hot channel temperature as a function of inserted reactivity, showing the effect of reactivity addition mode and Doppler feedback (LMFBR).

temperatures. Figure 2.13 demonstrates the difference between adding reactivity as a terminated ramp and as a step, and it also shows the very significant improvement which arises from an increase in the Doppler coefficient. Such an analysis defines the need for protective system response times of a given value depending upon the control rod malfunctions which are possible: if the control accident could add reactivity at a rate of $ 2/sec, then the protective system would need to cut back reactivity following a high flux signal in about 0.25 sec for example.

2.3.2.2 *In the Subthermal Range*

Here the system is not at significant temperature and so a rise in power produces no significant Doppler feedback to help to cut back the transient. Thus the power rise might progress through several decades in flux before significant feedback is induced. During this stage of the calculation only the neutron kinetics equations are needed.

The following trip signals are available: (a) control rod drive sensors; (b) period meters if they are included; (c) low flux; (d) intermediate flux; (e) high flux; and (f) high coolant temperatures eventually. Again, it is possible to define highest acceptable rates of reactivity addition if the protective system is well defined.

Figures 2.14–2.16 show power, fuel temperatures, and reactivity feed-

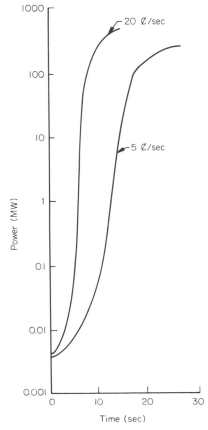

Fig. 2.14. The effect of continuous rod withdrawal at start-up. Power variations are shown as a function of time (LMFBR).

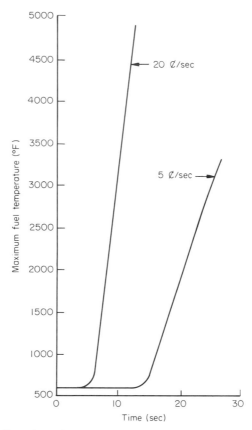

Fig. 2.15. The effect of continuous rod withdrawal at start-up. Maximum hot channel temperatures are shown as a function of time (LMFBR).

backs involved following a continuous rod withdrawal initiated at low power. Two rod withdrawal rates are shown for a typical LMFBR. In both cases even the fourth of the above sequence of trip signals will maintain acceptable conditions within the fuel. Figure 2.16 very clearly shows how important in each case the Doppler reactivity feedback is in reducing the reactivity addition and curtailing the power rise. In the 5¢/sec addition case, no feedback occurs for 15 sec, but when it does occur, the power is almost immediately curtailed.

2.3.2.3 Step Additions of Reactivity

Although control rod malfunctions cannot give rise to large step changes of reactivity, during any survey of acceptable reactivity changes, it is useful to include a parametric survey of the effects of large step changes. As will be seen, it is possible to envelope many other reactivity effects within an acceptable step value.

Figures 2.17 and 2.18 show power rises and temperature rises for a range of step additions which show that approximately 60¢ would be acceptable for the particular control system considered. It is important to realize that this acceptable value is dependent on the failure criterion chosen (here it is incipient fuel melting) and on the response of the protective system.

Many of the other postulated mechanisms for adding reactivity that follow will actually result in total reactivity changes that are less than the acceptable step value. Thus all can be shown to be acceptable without performing separate reactivity transients in each separate case.

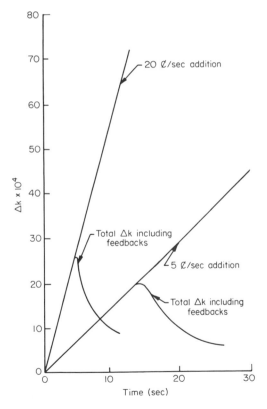

Fig. 2.16. The effect of continuous rod withdrawal at start-up, for different withdrawal rates. Reactivity variations are shown together with contributions due to feedback effects (LMFBR).

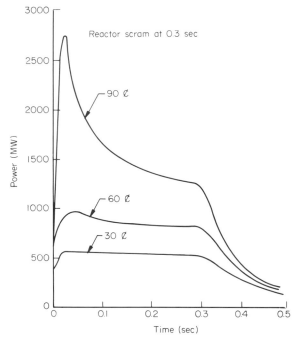

Fig. 2.17. The transient response of the reactor power level for different step additions of reactivity (LMFBR).

2.3.3 ADDITION OF MODERATOR

Those who work with fast reactors have to contend with the fact that an addition of moderator softens the spectrum and effectively increases the reactivity but only when this moderator is applied throughout the core. If the moderator is applied at the center, then the reactivity is liable to be reduced, whereas a reduction of moderator would cause an increase in reactivity. See Section 1.4.1.2 for a more detailed discussion of this point.

An LMFBR is not subject to moderator addition but to subtraction (Section 2.3.4). Gas-cooled and steam-cooled reactor systems are sensitive to the flooding accident, which is a moderator addition (2).

Figure 2.19 shows the variation in reactivity following the ingress of water as a result of a failure of heat exchanger tubes in the gas-cooled fast reactor system. An amount equal to 10 kgm/sec is assumed to leak in from a simultaneous failure of three tubes. However, in this case, the addition of reactivity is at a slow enough rate for detection to remedy the situation. Temperature changes in the fuel were about 0.2°C/sec.

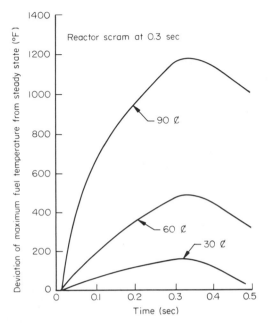

Fig. 2.18. The transient response of hot channel fuel temperatures for different step additions of reactivity (LMFBR).

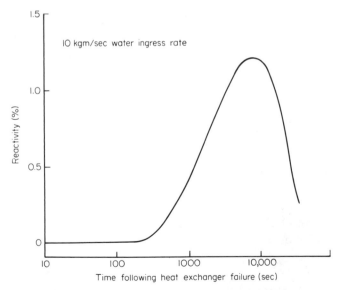

Fig. 2.19. The variation of reactivity in time following the failure of three heat exchanger tubes in a gas-cooled fast reactor allowing the ingress of 10 kgm of water per sec (2).

2.3.4 SUBTRACTION OF MODERATOR

As previously discussed, a voiding in the center of a LMFBR core results in a reactivity increase. There are two methods of subtracting moderator: (a) the introduction of bubbles, and (b) sodium boiling.

Boiling is dealt with in ensuing chapters of this book. It suffices to say here that if the coolant is boiling, then something is radically wrong before any reactivity change arises from the core voiding!

Bubbles could be introduced from a variety of sources such as entrained gases, fission product gases, chemical reaction products, and such sources are separately discussed in Section 5.4.1.

	Outlet 1	2	3	CORE ANNULUS NUMBER 4	5	6	7	8
16	-0.01	-0.05	-0.08	-0.05	-0.15	-0.15	-0.15	-0.15
15	-0.02	-0.10	-0.12	-0.05	-0.20	-0.27	-0.36	-0.40
14	-0.08	-0.37	-0.42	-0.10	-0.52	-0.93	-1.28	-1.50
13	-0.12	-0.72	-0.93	-0.27	-1.00	-1.77	-2.22	-2.60
12	-0.12	-0.54	-0.60	-0.62	0.02	-1.55	-3.30	-5.21
11	0.30	1.51	2.82	2.90	6.30	3.14	-1.10	-5.11
10	0.68	3.45	6.00	5.59	12.18	4.46	1.15	-4.8
9	0.90	4.56	7.86	7.18	15.63	10.00	2.45	-4.8
8	0.90	4.56	7.86	7.18	15.63	10.00	2.45	-4.8
7	0.68	3.45	6.00	5.59	12.18	4.46	1.15	-4.8
6	0.30	1.51	2.82	2.90	6.30	3.14	-1.10	-5.11
5	-0.12	-0.54	-0.60	-0.62	0.02	-1.55	-3.30	-5.21
4	-0.12	-0.72	-0.93	-0.27	-1.00	-1.77	-2.20	-2.60
3	0.08	-0.37	-0.42	-0.10	-0.52	-0.93	-1.28	-1.50
2	-0.02	-0.10	-0.12	-0.05	-0.20	-0.27	-0.36	-0.40
1	-0.01	-0.05	-0.08	-0.05	-0.15	-0.15	-0.15	-0.15

edge of positive voiding region* (rows 12/11)

CORE MIDPLANE (rows 9/8)

Inlet

*A totally voided positive void region is worth $2 in this case

Fig. 2.20. The sodium void reactivity worth as a function of core position in a LMFBR.

Figure 2.20 shows the worth map of sodium across a typical LMFBR core while Fig. 2.21 shows the variation of reactivity as a bubble passes through a core as a function of the bubble width, height, and coherency. These curves are calculated by simply integrating the worth map, and rates of reactivity addition are produced from an assumption as to the probable bubble size, shape, and velocity through the core. Rates of $100/sec are possible with investments of over a dollar but such sizes of bubbles are incredibly large.

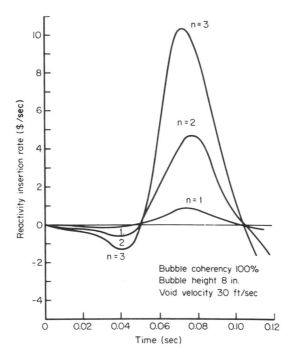

Fig. 2.21. The reactivity insertion rate as a function of time for a bubble distribution of a given length, *N* in. in diameter, moving at a given velocity through the core. The reactivity insertion rate is scaled by the coherence of the bubble distribution.

2.3.5 STRUCTURAL MOVEMENT

Such structural movements can arise as thermally induced expansion movement (as in EBR-I: Section 2.5.5) and through a thermal model they can be related to temperatures in the core. Otherwise, the structural movement may arise from suddenly released thermal restraint or from seismic forces that can cause a movement of fuel, fuel support, and/or control material.

A structural analysis based on equivalent spring-mass systems gives the accelerations of various components for given seismic frequencies. Pessimistic vibratory modes give maximum deflections from which reactivity changes can be calculated on a steady-state basis. This can be done for vertical movement or radial movement of the fuel assembly considered as a simple supported beam. Such calculations will give pessimistic values since the fuel pins will be well supported by grids or wire wrap spacers and assemblies are usually close-packed.

Typical seismic movements might be in the range 0.01–0.1 in. resulting in reactivity changes of up to 10¢ and generally well within acceptable reactivity steps for adequate system protection to be guaranteed. These considerations apply equally to all three systems under discussion.

2.4 Thermal Perturbations

2.4.1 SYSTEM MODELING

Section 1.3 showed that thermal effects formed a link between the power input to the system and the feedback reactivity effects which, in turn, affect power. Thermal modeling was included in the section.

Thermal changes in the system can be system-induced but they can also be induced by external means: (a) inactive loop start-up, (b) heat exchanger rupture, (c) secondary pump failures, (d) feed supply failure, (e) turbine stop-valve closure, or a large number of minor steam cycle malfunctions.

These disturbances will be witnessed by the core as a change in the inlet temperature. This has, in turn, three effects: (a) an immediate coolant reactivity temperature feedback; (b) a delayed Doppler feedback; and (c) a further delayed (by the loop time constant) effect on the inlet temperature.

All these effects may be taken into account in a comprehensive system model such as would describe Fig. 1.20. Some models which do not represent the steam cycle may have to be supplemented by additional calculations performed to produce a perturbation for the model in the form of the inlet temperature as a function of time.

Trips available during such transients are the following: (a) primary signals such as pump control or valve control monitors or steam generator pressure relief signals; (b) thermocouples in secondary and tertiary loops; (c) heat exchanger primary outlet temperature; and (d) core outlet temperature (eventually). There is considerable time available in steam cycle incidents due to the insulation of the core from the steam cycle by the intermediate or secondary loop. The trips are in order of occurrence. It can be seen that the core is very well protected from this kind of disturbance.

2.4.2 SECONDARY PUMP FAILURES

These are treated in exactly the same way as primary pump coast-downs or seizures were, and the resulting flow changes are input to the model, which includes the secondary circuit details.

The resulting transients are, of course, much less serious than primary pump failures because of the attenuation by the secondary and cold leg primary sodium. They are more of interest in the operational sense than in the safety sense.

2.4.3 TERTIARY LOOP FAILURES

This failure is again one step further removed from the reactor core.

2.4.3.1 *LMFBR*

The tertiary loop in a power-producing LMFBR is the steam cycle, but in an experimental reactor such as FFTF (6), a bank of blowers constitutes the tertiary heat dump.

In a power-producing LMFBR there may be only a single steam cycle and feed loop even though the primary loop may be split into three. In this case a tertiary failure (feed-valve closure or failure of feed supply for example) would affect all the primary loops in the same way. However in an experimental system each of the primary and secondary loop systems can have its own heat dump air blast heat exchangers and then a failure in one of these would only affect a single primary and secondary loop combination.

In any case there are considerable delays (two cold leg circuit times) of the order of 20–30 sec before any tertiary loop disturbance is felt by the LMFBR core. The only requirement is that one fault should not remove *all* heat removal capability from the system, since even when the core is shut down successfully, the decay heat must still be removed.

The reactor can best be tripped from an intermediate heat-exchanger high primary outlet signal, although signals would available during the transient from the following sources: (a) tertiary component signals such as electrical supply circuits for loss of electrical supply initiators or pump speed indicators for a failure of the feed pump; (b) IHX primary outlet temperature; and (c) core outlet temperature.

Figure 2.22 shows assumed flows in primary, secondary, and tertiary loops which might arise if a LMFBR lost feed supply. Figure 2.23 shows typical steam generator sodium outlet temperatures that might arise from this failure together with IHX secondary temperatures and the IHX primary outlet temperature which eventually initiates a trip at about 8 sec. Figure 2.24 shows the reaction of primary and core temperatures. In fact the first indication in the core that a failure has occurred is that the system shuts down; no excess temperatures are recorded.

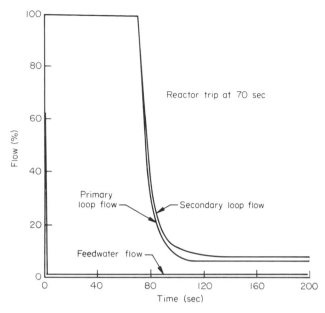

Fig. 2.22. Flows within a LMFBR system following a loss of feedwater supply followed by a reactor trip and pump trips on a high IHX primary outlet temperature signal.

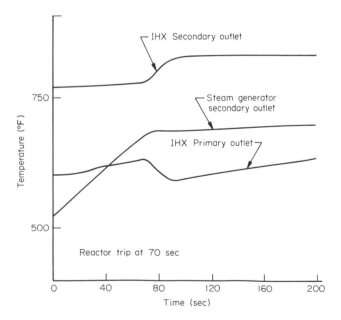

Fig. 2.23. Temperatures within the secondary and tertiary loops following a loss of feedwater supply and subsequent trips.

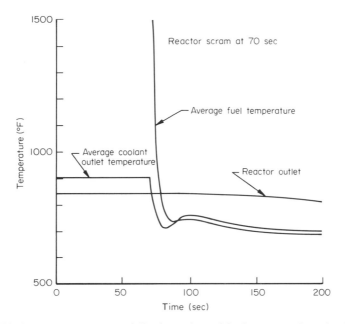

Fig. 2.24. Reactor temperatures following a loss of feedwater supply and subsequent reactor and pump trips.

2.4.3.2 *Steam-Cooled Fast Reactor*

This system may be a direct cycle with one or more stages of reheat and therefore a steam cycle fault is not a true tertiary loop fault. The effect is not a primary temperature increase but a primary pressure increase, because the steam cycle is, in a sense, part of the primary loop (Fig. 2.3). A throttle closure is a pressurization transient and is therefore in the opposite direction but much milder of course than the depressurization accidents of Section 2.2.4.3.

In Section 1.5.2.1 the load-following characteristics of a PWR were described. Load-following is possible because a primary thermal change is indirectly introduced by the operation of a steam cycle change (manipulation of the turbine stop-valve) (Fig. 1.21) and this gives rise to the correct reactivity change if the temperature coefficient is negative. Short circuit times allow load-following to be efficiently rapid in a PWR. This load-following is not possible in a LMFBR because of the negative central density coefficients and long delays in the intermediate liquid-metal loops. Temperature changes would be too large and in the wrong direction for a correct power change.

However load-following of a sort is possible in the steam-cooled fast reactor through its reheat loops. It is in fact a semi-indirect cycle in this respect. But in this case the primary core must be maintained and manipulated in unison with the turbine steam flow to achieve partial load-following. Other plant components such as the feed-valve operation will have a marked effect on the ensuing transient.

2.4.4 LOCAL THERMAL PERTURBATIONS

Local temperature and heat transfer changes in the LMFBR core will be treated in detail in Section 4.4. One possible perturbation not treated is the addition of cold sodium to the core. This might be presumed pessimistically to occur if an inactive loop is activated in some sudden fashion resulting in a reduction of the inlet temperature. The result would be a slight addition of reactivity if the overall coolant temperature coefficient were negative, which might be followed by a further reactivity addition as the fuel temperature was reduced. However, even on an instantaneous basis assuming very pessimistic inlet temperature changes, this reactivity change cannot be larger than that which the protective system can handle. In practice the system would be surveyed for possible ways in which the inlet could experience colder sodium flow, and the reactivity change would be calculated. It could then be shown to be small compared to the largest acceptable step of reactivity as defined in Section 2.3.2.3.

One further local temperature change not yet noted concerns the steam-cooled fast reactor system and the gas-cooled system. In reactivity or flow transients which result in coolant temperature variations, the coolant density is very sensitive to these temperature changes, and additional heat transfer variations will occur in addition to the usual reactivity feedbacks. Such effects will be included in the modeling by ensuring that the density of the coolant is allowed a pressure and temperature dependence.

$$m_{\mathrm{c}} = A\varrho_{\mathrm{c}} = A\varrho_{\mathrm{c}0}\,T_0 P/TP_0 \tag{2.7}$$

2.5 System Stability

In addition to its response to particular disturbances, the system must be assessed for its response to any general disturbance which might induce a core or plant instability.

2.5.1 GENERAL DISCUSSION

In Section 1.5.3, the mechanisms of instability that could occur were discussed. They have been associated with some voidage and in the main with water-cooled systems: they were characterized by having a combination of time constants and feedbacks which interacted in a deleterious manner.

The same can happen in principle in any system, so that any fast reactor design needs to be checked for stability before proceeding with the design. If it is not stable, then suitable modifications may be required to ensure stability.

Such modifications may either change the magnitude of a feedback or alter the time constant of that feedback. A nuclear reactivity coefficient may be given a certain value by the nuclear design of the core, or the flow paths may be given certain time constants by the addition of an additional pressure drop, or a reduction in piping length, or the alteration of plenum design.

First a summary of methods available for the assessment of system stability will be given.

2.5.2 SIMULATION TECHNIQUES

By preparing a model of the system and by perturbing this simulation with sample disturbances, it is possible to get a limited idea of the system stability. The weakness of this method is that it is impossible to cover all cases of perturbations and initial states (upon which feedbacks depend) and thus only a partial assessment is possible. The method therefore demands a critical choice of the power levels and instability modes and even the system components that might be of direct interest. A good deal of experience on the part of the evaluator is necessary.

The strength of the simulation technique is that quicker information is available about the effect of various delays, pressure drops and design changes if an analog simulation is used. For example, it is much easier to include a bypass line in an analog patch than it is in a transfer function analysis.

2.5.2.1 *Method*

If it were supposed that a given type of instability might occur in the system as a result of a change in a particular variable X_1, and that a design modification in another parameter X_2 could correct the situation, then the following method of assessing this instability could be used:

(a) Set up a mathematical simulation of the system to include the full dependence of both the system variable X_1 and the parameter X_2. (X_1 may be the power-to-flow ratio and the parameter X_2 may be a given feedback coefficient.)

(b) Disturb the model by a "kick" in a significant variable (say, pressure or flow) and observe the dynamic results. Figure 2.25 shows what the results might indicate.

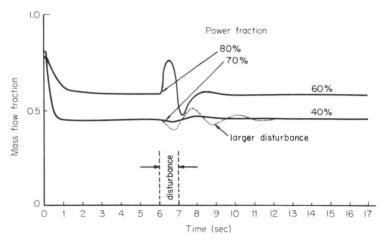

Fig. 2.25. Mass flow response to coolant flow disturbances in a stability investigation. The response is given as a function of the power and mass flow level.

(c) Calculate a damping factor from successive peaks of the transient.

$$\text{damping factor } \lambda = (x_2 - x_1)^{-1} \ln(y_1/y_2) \tag{2.8}$$

where y_i and x_i are defined in Fig. 2.26.

(d) Vary sensitive parameters for their effect on the damping factor; this sensitivity study would include both the system variable X_1 and the design parameter X_2. Figure 2.27 shows the result of such a sensitivity study. [The damping factor is a system characteristic and does not depend on the variable used for its calculation (mass flow shown in Fig. 2.25).]

(e) Invoke the design parameter needed to achieve the damping factor required. In Fig. 2.27, damping factors of 2 and above are acceptable while

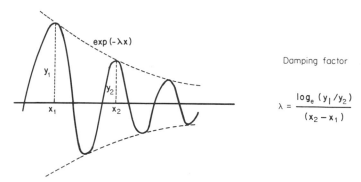

Damping factor

$$\lambda = \frac{\log_e (y_1/y_2)}{(x_2 - x_1)}$$

Fig. 2.26. A geometric definition of the damping factor.

those below about 1.5 show poor damping, neutral stability without damping, and finally instability. Thus, if X_1 (say the power-to-flow ratio) can be
1 in this system, then X_2 (the feedback coefficient) must be designed to be
at least B.

This method can achieve very rapid results if an analog computation is used.
Digital methods, which may be more comprehensive, can then be used to
check the result.

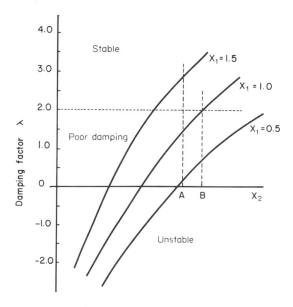

Fig. 2.27. Damping factor of a reactor system as a function of two system characteristics: X_1 (say a power-to-flow ratio) and X_2 (say a feedback coefficient).

2.5.3 Transfer Function Analysis

All the systems to be assessed are closed-loop feedback systems.

A simple dynamic system (Fig. 2.28a), which has a response R to an input signal I, can be represented by a transfer function $G(s) = R/I$, where s is the Laplace variable, which is generally a complex variable. The response of this system to a steady sinusoidal input of frequency ω of unit amplitude is $R = G(i\omega)$, which is called the frequency response (7).

Fig. 2.28a. Simple dynamic system.

A simple feedback loop (Fig. 2.28b) has a forward function $G(s)$, a feedback transfer function $H(s)$, and a feedback signal F, which is the output R modified by the feedback function

$$F = RH(s) \tag{2.9}$$

The input to the forward transfer function $G(s)$ is now the difference between the input signal I and the feedback F. It is clear that with the loop closed the response R is given by

$$R = G(s)[I - RH(s)] \tag{2.10}$$

$$R/I = G(s)/[1 + G(s)H(s)] \tag{2.11}$$

and it should be noticed that the positive sign in the denominator is indicative of a negative feedback.

This is called the *closed-loop transfer function*.

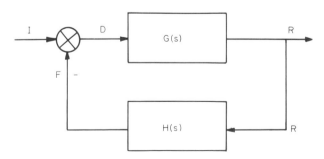

Fig. 2.28b. Simple feedback dynamic system.

The function $G(s)H(s)$, the feedback response to unit amplitude F/I, lis called the *open-loop transfer function*.

Instability in the system is exhibited when the signals D, R, and F in the oop become self-sustaining without an input I. Instability is indicated by the poles (where the function becomes infinite) of the closed-loop transfer function in the right half of the complex plane s, where the poles indicate exponentially increasing time functions in the time domain.

The number of poles is equal to the number of unstable modes in the system while the position of the pole gives information about the type of instability shown; the real s coordinate is the divergence rate while the imaginary s coordinate is the divergence frequency. There are several methods by which this information can be used to analyze the stability of the reactor system.

2.5.3.1 *Location of Poles*

This method does exactly what it states, it locates the poles by solving the characteristic equation from the *closed-loop* transfer function:

$$1 + G(s)H(s) = 0 \qquad (2.12)$$

In a reactor system, these poles will be functions of core parameters, the power level, and plant time constants.

Depending on the degree of s in $G(s)H(s)$, various criteria can be established for defining the stability of the system. These are algebraic conditions on the coefficients of Eq. (2.12) to ensure that there are no poles in the right half of the complex s plane.

a. *Routh–Hurwitz method.* This method (7) starts from the closed-loop characteristic [Eq. (2.12)], written as

$$a_n s^n + a_{n-1} s^{n-1} + \cdots + a_0 = 0 \qquad (2.13)$$

Then an array of $n + 1$ rows is prepared as follows:

$$
\begin{array}{llll}
a_n & a_{n-2} & a_{n-4} & \cdots \\
a_{n-1} & a_{n-3} & a_{n-5} & \cdots \\
b_1 & b_2 & b_3 & \cdots \\
c_1 & c_2 & c_3 & \cdots, \text{ etc.}
\end{array} \qquad (2.14)
$$

where

$$b_1 = (a_{n-1}a_{n-2} - a_n a_{n-3})/a_{n-1}$$

$$b_2 = (a_{n-1}a_{n-4} - a_n a_{n-5})/a_{n-1}, \quad \text{etc.}$$

(2.15)

and

$$c_1 = (b_1 a_{n-3} - b_2 a_{n-1})/b_1, \quad \text{etc.}$$

The number of roots is the number of sign changes in the first column of this array (2.14). Thus the stability criterion is that there should be *no* changes of sign in this column and thus no poles in Eq. (2.11).

b. *Root-Locus method.* This method also ensures no poles in the right half of the plane by drawing a locus of values of s which satisfies the characteristic Eq. (2.12). The locus is drawn by a graphical method and then the stability criterion is stated in terms of the points at which the locus passes into the right half-plane. A knowledge of $G(s)H(s)$ is required as well as its zeros and poles and the locus is drawn from these points such that a value of s on the locus satisfies the two equations

$$| G(s)H(s) | = 1$$

$$\arg G(s)H(s) = n\pi$$

(2.16)

which is Eq. (2.12) in its gain and phase components. The rules for drawing the root-locus and the statement of the criteria for stability are summarized in Table 2.2. The subject is treated in excellent fashion by Weaver (7).

c. *Solution of the characteristic equation.* If the system is simple, then one final method is available. The characteristic equation could be solved for its roots [that is, for the poles of Eq. (2.11)], given all the values of the relevant coefficients, which would include heat-transfer coefficients, feedback coefficients, and time constants. In even a simple system, this method can be, at best, time-consuming; in a more complex system it is generally impossible.

2.5.3.2 *Frequency Response Analysis*

This is an alternative analytical method in which the stability is assessed from an *open-loop* analysis.

The frequency response of the open-loop system $G(s)H(s)$ is plotted in two distinct ways to give two equivalent stability criteria.

a. *Nyquist criterion.* This criterion allows one to calculate the gain and phase margins to give the degree of stability of the system, whereas the

TABLE 2.2

PLOTTING RULES FOR ROOT-LOCUS CONSTRUCTION

Step	Plotting rule
1	Number of locus branches equals order of denominator of $G(s) H(s)$.
2	Locus is drawn for values of K as K varies from zero to infinity where K is given by $$G(s) H(s) = K (s - z) \cdots /(s - p) \cdots.$$
3	As K tends to infinity the locus of the closed-loop poles terminates on a zero of $G(s) H(s)$. Zeros at infinity are included. Also as K tends to zero the locus approaches the poles of $G(s) H(s)$.
4	Loci on the real axis include those sections of the axis to the left of an odd number of critical frequencies (zeros or poles) of $G(s) H(s)$.
5	Loci near infinity asymptotically approach straight lines which meet the real axis at an intersection point given by $$\frac{\text{sum(real part of poles)} - \text{sum(real part of zeros)}}{\text{number of finite poles } (p) - \text{number of finite zeros } (z)}$$ and they meet the axis at angles given by $\theta = n\pi/(p - z)$ for n odd.
6	Loci have symmetry about the real axis.
7	The values of K at which loci go into the right half-plane can be determined graphically.

location of poles allowed one to obtain a rate of exponential decay that could be related to a damping factor as in the simulation technique of Section 2.5.2.1.

The frequency response of the open-loop system $G(s)H(s)$ is plotted on an Argand diagram and the criterion states: "If a system is unstable the number of unstable frequencies is given by the number of times this frequency response curve encircles the -1 point in a clockwise direction."

Defining

$$W(s) = G(s)H(s) + 1 \qquad (2.17)$$

to be the closed-loop characteristic, from Cauchy's theorem in complex number theory, assuming that $W(s)$ is single-valued and has finite poles in a contour C and is not zero on C, then

$$\int_C W'(s)/W(s) = 2\pi i(Z - P) \qquad (2.18)$$

so

$$[\ln W(s)]_C = 2\pi i(Z - P) \tag{2.19}$$

$$= A_2 + iB_2 - A_1 - iB_1 \tag{2.20}$$

where Fig. 2.29a defines the anticlockwise path around contour C.

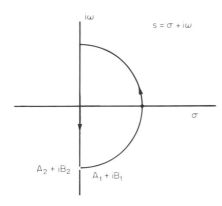

Fig. 2.29a. C contour in the s plane.

Thus

$$A_1 = A_2 \tag{2.21}$$

$$B_2 - B_1 = 2\pi(Z - P) \tag{2.22}$$

One revolution of the W plane origin gives $2\pi i$, thus the number of encirclements of the W origin is equal to $Z - P$.

The stability criterion now arises because the closed-loop transfer function pole is a zero of $W(s)$ and therefore for stability there should be no zeros in the right half-plane of $W(s)$. For stability there must be as many encirclements as $W(s)$ has poles in the right half-plane.

Thus the number of encirclements of the origin in the $W(s)$ plane $(Z - 0)$ is equal to the number of encirclements of the point -1 in the $G(s)H(s)$ plane, and this in turn is equal to the number of poles of the open-loop transfer function as shown above. The negative sign now implies clockwise encirclements. Examples are given in Section 2.5.3.3.

The effect of a time lag is to make the closed-loop system more unstable in general. See the second Nyquist example in Section 2.5.3.3.

b. *Bode plot method.* Again using the open-loop response of the system, the phase and gain may be plotted and the Nyquist criterion used in a slightly different manner.

The gain is defined as M where:

$$M = 20 \log_{10} \quad \text{(response argument)} \tag{2.23}$$

and the gain and phase are plotted against frequency. Because of the logarithmic formulation the gains and phases can be added for multiplicative systems, so that the frequency responses of sections of the system can be easily totalled. See the example in Section 2.5.3.3.

Thus the response of the neutron kinetics can be calculated separately from the thermal characteristic response, and the feedback and each can be converted into a separate tabulation of gain and phase versus frequency which can be summed. The only diffculty occurs where multiple feedback paths exist.

The Nyquist diagram for the -1 point is translated here into a similar one on the Bode diagram for the point of 0 dB and a phase of $-180°$. For mere stability, the system must have a negative gain at $180°$ but, in practice, for suitable operational stability one needs some damping, which is provided by a gain margin of about 20 dB (Fig. 2.29b).

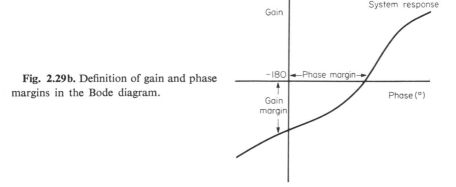

Fig. 2.29b. Definition of gain and phase margins in the Bode diagram.

2.5.3.3 Examples

Several examples are presented to clarify certain sections of the foregoing text

a. *Routh–Hurwitz criterion* (refer to Section 2.5.3.1). Figure 2.30 shows a simple feedback loop for an imaginary fluid-fueled system in which the fuel produces a power P while in the critical configuration of the core but gives up a proportion of its heat a while on a path through an IHX. The circuit time is θ. The equations are:

Fuel

$$\partial T/\partial t = P - h(T - T_i) \tag{2.24}$$

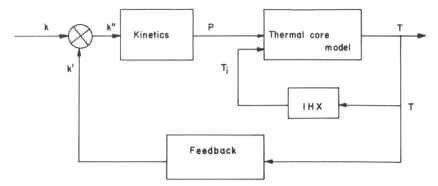

Fig. 2.30. The primary loop feedback system for an imaginary fluid-fueled reactor.

Neutron kinetics

$$\partial P/\partial t = (\delta k\, P/l^*) - (\alpha T/l^*) + (\alpha T_0/l^*) \tag{2.25}$$

Circuit

$$T_i = aT \qquad \text{with a delay of} \quad \theta \tag{2.26}$$

No delayed neutrons are assumed. Taking the Laplace transform,

$$sT - T_0 = P - hT + hT_i \tag{2.27}$$

$$sP - P_0 = (\delta k\, P_0/l^*) + (\delta k_0\, P/l^*) - (\alpha T/l^*) \tag{2.28}$$

$$T_i = aT/(1 + \theta s) \tag{2.29}$$

and in the steady state it is assumed that

$$T_{i0} = aT_0 \qquad \text{and} \qquad \delta k_0 = 0$$

$$P_0 = h(T_0 - T_{i0}) \qquad \text{and take} \qquad P_0 = 1 \tag{2.30}$$

From these relations the frequency response of each pair of variables $T/\delta k$, $P/\delta k$, and $T_i/\delta k$ may be calculated. Defining $G(s) = P/\delta k''$ and $H(s) = \delta k'/P$, where

$$\delta k'' = \delta k - \delta k' \qquad \text{and} \qquad \delta k' = (\alpha T - \alpha T_0) \tag{2.31}$$

the characteristic equation is given by

$$G(s) = \frac{1}{sl^*} \tag{2.32}$$

$$H(s) = \frac{\alpha(1 + s\theta)}{s^2\theta + s(1 + \theta h) + h(1 + a)} \tag{2.33}$$

so $1 + G(s)H(s) = 0$ gives a cubic equation:

$$s^3 l^* \theta + s^2 l^* (1 + \theta h) + s(l^* h(1 + a) + \alpha \theta) + \alpha = 0 \qquad (2.34)$$

The Routh–Hurwitz criterion for stability gives the following array:

$$
\begin{array}{ll}
l^*\theta & l^*h(1+a) + \alpha\theta \\[1em]
l^*(1+\theta h) & \alpha \\[1em]
l^*(1+a)h + \dfrac{h\alpha\theta^2}{(1+\theta h)} & 0 \\[1em]
\alpha & 0
\end{array} \qquad (2.35)
$$

from which the criterion of no sign changes in the first column gives:

$$\alpha > 0 \quad \text{and} \quad l^*h(1+a) + \frac{\theta^2 h\alpha}{(1+\theta h)} > 0 \qquad (2.36)$$

Both of these statements are satisfied by a positive α, so the system is stable for all negative Doppler coefficients $(-\alpha)$.

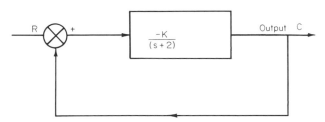

Fig. 2.31. Closed-loop system (7) for $G(s) = -K/(s+2)$.

b. *Nyquist criterion.* (1) A system in which $G(s) = -K(s+2)^{-1}$ is shown in Fig. 2.31. To draw the locus of $G(i\omega)$ notice that for

$$
\begin{aligned}
\omega = 0, \quad & G(i\omega) = -\tfrac{1}{2}K \\
\omega = \infty, \quad & G(i\omega) = 0
\end{aligned} \qquad (2.37)
$$

Thus the locus starts from the origin and has a diameter of $\frac{1}{2}K$ (Fig. 2.32). The number of poles P of $G(s)$ in the right half-plane is zero (only $s = -2$ exists in the left half-plane) so, for stability $(Z = 0)$, there should be no encirclements of the origin by $1 + G(s)$. To avoid any encirclements by $G(s)$ of the point -1 and to ensure stability for this system, the criterion is that K must be less than 2. Any adjustment to the system to make K smaller to obtain this condition will be a gain adjustment.

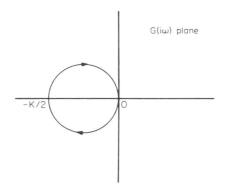

G(iω) plane

−K/2

O

Fig. 2.32. Nyquist plot (7) for $G(i\omega)$.

(2) If the system is now represented by

$$G(s) = Ks^{-1}(s + a)^{-1}(s + b)^{-1} \tag{2.38}$$

with a feedback loop shown in Fig. 2.33, the locus in the $G(i\omega)$ plane is shown in Fig. 2.34a as for

$$\begin{aligned} \omega = 0, &\qquad G(i\omega) = \infty \\ \omega = \infty, &\qquad G(i\omega) = 0 \end{aligned} \tag{2.39}$$

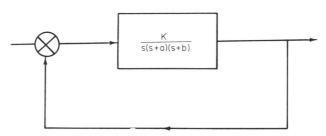

$$\frac{K}{s(s+a)(s+b)}$$

Fig. 2.33. Closed-loop system (7) for $G(s) = K/[s(s + a)(s + b)]$.

The plot in the s plane is shown in Fig. 2.34b. Here $P = 0$ for $a,b > 0$ so there should be no encirclements. The stability criterion is now that $K/ab(a + b)$ must be less than unity to avoid encirclements of the point -1. Thus the maximum gain K is $ab(a + b)$. This is called the gain margin when related to the actual value of K.

Note that in this example an nth order pole at the zero point in the s plane is mapped into a counterclockwise rotation of $n\pi$ of infinite magnitude in the $G(s)$ plane. $H(s)$ is, of course, unity in these examples.

(3) The effect of a time lag is that it tends in general to make the closed-loop system more unstable. The lag $\exp(-sT)$ in association with $G(s){=}K/s$

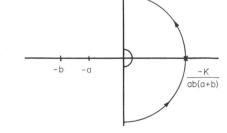

Fig. 2.34a. The s plane contour (7).

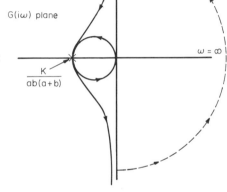

Fig. 2.34b. Nyquist diagram (7) for
$G(i\omega) = K/[i\omega(i\omega + a)(i\omega + b)]$.

gives the Nyquist diagram shown in Fig. 2.35, where the time lag is a linearly increasing, frequency-dependent function. For this system, the stability criterion is that K should be less than $\frac{1}{2}\pi/T$.

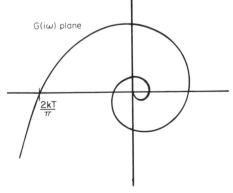

Fig. 2.35. Nyquist plot for systems with time delay (7).

c. *Bode plot.* For a system characterized by

$$G(s) = K/(1 + sT)^n \qquad (2.40)$$

the gain is

$$20 \log| G(i\omega) | = 20 \log K - n20 \log| (1 + iT\omega) | \qquad (2.41)$$

The separate parts of this expression can be plotted and added as shown in Fig. 2.36.

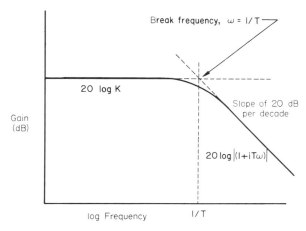

Fig. 2.36. Log magnitude as a function of the log of the frequency.

If now a feedback $H(s) = F(1 + s\theta)$ is required, then the gain of $G(s)H(s)$ is given by the sum

$$\text{total gain} = 20 \log K - 20n \log|(1 + iT\omega)| + 20 \log F + 20 \log|(1 + i\theta\omega)| \qquad (2.42)$$

for all frequencies ω. In order to establish the stability margin, this gain is plotted against the phase after eliminating the frequency parameter. The criterion of Section 2.5.3.2 is used.

2.5.4 NONLINEARITIES

All the foregoing text has referred to linear stability analysis and any system with nonlinearities would have to be linearized before being treated in this way. [Witness Eq. (2.28), which linearizes the $\delta k\, P$ term of Eq. (2.25).] However, many systems are significantly nonlinear.

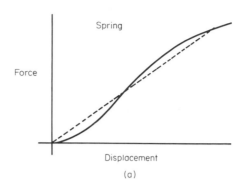

Fig. 2.37. (a) Nonlinear system: spring displacement for a given force. (b) Nonlinear system: saturated system behavior.

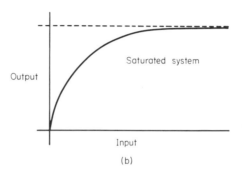

Figure 2.37a and b illustrates (a) the nonlinear behavior of a system composed of a force acting on a spring and its linearized representation, and (b) a system that saturates for a high enough input, which can clearly be linearized only for small changes or for high input values.

There are nonlinear stability criteria, but Liapunov's theorem states that linear analysis is valid for small disturbances even in nonlinear systems.

The simulation technique is a nonlinear method if the simulation includes the nonlinear terms. The ideal initial assessment would include a simulation technique in conjunction with other methods to check on their validity before delving further into nonlinear methods of analysis and their corresponding criteria for stability (*8, 9*).

2.5.5 EXPERIENCE

Apart from Fermi steam generator problems (Section 4.6), the only fast reactor which has exhibited a substantial instability is the original core of EBR-I. This proved to be the result of a secondary fuel rod bowing effect, by which a primary prompt positive reactivity change, produced by an original bowing of the fuel rods, was counteracted by a delayed and negative effect (*10*).

When the system was brought to power, the fuel rod first bowed inward while supported by the first shield plate and the lower grid plate, but some time later the rod began to be affected by the expansion of the third shield plate which then effectively moved the fuel element away from the center of the core giving a slow negative reactivity coefficient for large power-to-flow ratios (Fig. 2.38).

The diagnosis of this effect was very difficult but the problem was solved eventually by using a restrained core to prohibit all bowing, including the

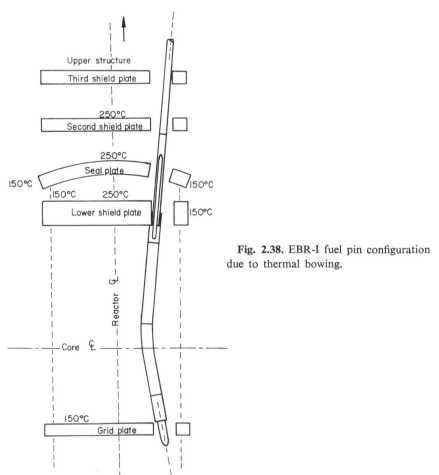

Fig. 2.38. EBR-I fuel pin configuration due to thermal bowing.

original prompt positive effect. The absence of any instability confirmed the analysis.

2.5.5.1 *Fermi Stability Analysis*

The Enrico Fermi reactor, designed during the time when the EBR-I instability analysis was underway, had considerable attention paid to its stability characteristics and it was designed with a very restrained core. It has experienced no adverse primary instability effects. The stability analysis of the reactor was particularly complete (*11*). It is here reproduced in full as an example of fast reactor stability analysis.[†]

The heat transport system is a purely passive system, and any instability would have to have its origin in the reactor through an adverse coupling of the thermal reactivity feedback with the neutron kinetics. Feedback can be considered as the sum of two components: one, the internal feedback, caused directly by a power change, assuming a constant reactor inlet coolant temperature, and the other, the external feedback, caused by a change of the reactor inlet coolant temperature fed back around the coolant loops from a change of the reactor outlet coolant temperature. The relations involved in the kinetics of the reactor, including the feedback, can be represented schematically by the signal flow diagram of Fig. 2.39. The variables corresponding to each node are listed on

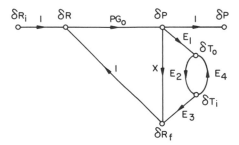

Fig. 2.39. Reactor kinetics signal flow diagram (*10*). R_i = inserted reactivity; R_f = feedback reactivity; R = total reactivity $R_i + R_f$; P = reactor power; T_o = reactor outlet coolant temperature; T_i = reactor inlet coolant temperature; $G_0 = \delta P/P \, \delta R$ = zero power reactor transfer function; X = power coefficient of reactivity, assuming constant T_i; $E_1 = \delta T_o/\delta P$, assuming constant T_i; $E_2 = \delta T_i/\delta T_o$ = the transfer function for the transmission of a temperature signal around the coolant loops (part of this transmission is around the primary loop only, and part around the primary and secondary loops in series); $E_3 = \delta R_f/\delta T_1$, assuming constant power; $E_4 = \delta T_o/\delta T_1$, assuming constant power in particular; $E_3(0)$ = isothermal temperature coefficient of reactivity; and $E_4(0) = 1$.

[†] Equation, reference, and figure numbers referred to in quoted material here and on following pages follow sequence of the present volume, not the text of the original.

the figure. Each branch linking two nodes represents the transmittance or transfer function linking the corresponding variables.

By using conventional techniques, the reactor transfer function G_p at power P is:

$$G_p = \frac{\delta P/P}{\delta R_i} = \frac{G_0}{1 - PG_0\{X + [E_1 E_2 E_3/(1 - E_2 E_4)]\}} \tag{2.43}$$

One can define a total power coefficient X_t:

$$X_t = X + \frac{E_1 E_2 E_3}{1 - E_2 E_4}$$

so that

$$G_p = \frac{G_0}{1 - GP_0 X_t} \tag{2.44}$$

The feedback appears as a sum of two components acting in parallel. Hereafter X will be called the internal power coefficient (or simply the power coefficient) and $E_1 E_2 E_3/(1 - E_2 E_4)$ the external power coefficient.

It is well known that the system will be stable if $P < P_c$, where P_c is the minimum power for which G_p has a pure imaginary pole $i\omega_c$; this can be expressed mathematically as

$$1 - P_c G_0(i\omega_c) X_t(i\omega_c) = 0 \tag{2.45}$$

At $P = P_c$, the reactor would exhibit an oscillatory instability at angular frequency ω_c.

Following construction of the reactor oscillator, tests run at progressively higher powers on the reactor will give a clear indication of whether or not any such power P_c exists.

If it is assumed that the feedback is not significantly different from what has been calculated, it is possible to demonstrate, using in part results obtained by simulation, that no instability is possible. At frequencies higher than a certain value f_1, the external feedback is for practical purposes completely attenuated and can thus be neglected. At frequencies lower than a certain value f_2, all temperatures inside the reactor respond to power or inlet coolant temperature changes in a quasi-steady state fashion, and hence X, E_1, E_3, E_4 are equal to their steady state values $X(0)$, $E_1(0)$, $E_3(0)$, $E_4(0)$. From simulation studies it was found that f_1 and f_2 are coincident and have the value of 0.01 cps. Since $f_2 = f_1$, the entire frequency range can be covered by the ranges $f > f_1$, and $f < f_2$, with the two ranges meeting at $f_1 = f_2$.

According to the definitions of f_1 and f_2, the total power coefficient $X_t = X + [E_1 E_2 E_3/(1 - E_2 E_4)]$ can be simplified as follows:

$$X_t = \begin{cases} X & \text{for } f > f_1 \tag{2.46} \\[2mm] X(0) + \dfrac{E_1(0)E_2 E_3(0)}{1 - E_2 E_4(0)} & \text{for } f < f_2 \tag{2.47} \end{cases}$$

Since the two frequency ranges cover all frequencies, it is sufficient to investigate separately whether instability is possible in either of the two ranges.

Stability analysis in the higher frequency range. In the frequency range $f > f_1$, $X_t = X$. The stability criterion given by Eq. (2.45) can be expressed in terms of amplitude and phase, with the feedback separate from the neutron characteristics:

$$\text{Amplitude:} \quad P \mid X(i\omega_c) \mid = 1/|G_0(i\omega_c)| \tag{2.48}$$

and

$$\text{Phase:} \quad \phi_X(i\omega_c) = \pi - \phi_{G_0}(i\omega_c) \tag{2.49}$$

where ϕ_X and ϕ_{G_0} are the phase lags of X and G_0, respectively. Instead of Eq. (2.49), the following expression may be used.

$$\tau_b(i\omega_c) = \phi_X/\omega_c = [\pi - \phi_{G_0}(i\omega_c)]/\omega_c \tag{2.50}$$

At any frequency ω, the phase lags of X and G_0 can be expressed as functions of $\mid X \mid$ and $\mid G_0 \mid$ respectively as

$$[\pi - \phi_{G_0}(i\omega)]/\omega = f(1/|G_0(i\omega)|) \tag{2.51}$$

and

$$\tau_b(i\omega) = \phi_X(i\omega)/\omega = g(P \mid X(i\omega) \mid) \tag{2.52}$$

The functions f and g are single valued explicit functions; however, in practice they are usually treated as implicit functions of the variable ω.

Equation (2.51) is dependent only on the neutron kinetics characteristics of the system and Eq. (2.52) is dependent only on the feedback reactivity characteristics of the system. At the critical point for instability, $\omega = \omega_c$, $f = g$, and the curves for these two equations intersect. Hence Eqs. (2.51) and (2.52) are equivalent to Eqs. (2.48) and (2.50) at the critical point. For any reactor, Eq. (2.51) can be calculated as a function of the argument with good accuracy, since it involves only neutron kinetics parameters. This equation is represented on Fig. 2.40 for the Fermi reactor, and also for a reactor with a neutron lifetime of 10^{-4} sec. The two curves diverge only for frequencies beyond the range where instability is most likely. (At high frequencies where $f \gg f_1$ the amplitude X of the feedback is so attenuated and $1/G_0$ is so large that instability is not likely.) Since the dollar was chosen as the unit of reactivity, the intermediate part of the curve, where the abscissa is unity, would not be affected by a different value of the delayed neutron fraction. At low frequencies the curve would only be slightly affected by the different delayed neutron characteristics of other fissionable isotopes. Hence for all practical purposes the neutron kinetics curve of Fig. 2.40 is almost universal regardless of the reactor type of interest.

In order to investigate whether or not a reactor is stable at power P, there is also plotted on Fig. 2.40 the feedback reactivity curve with coordinates $P \mid X(i\omega) \mid$ and $\tau_b(i\omega)$, when ω is varied from zero to infinity. This curve should be analyzed in relation to the neutron kinetics characteristics curve shown on Fig. 2.40. If at some value of ω, the points of both curves have the same ordinate, relation (2.50) is satisfied, and the phase is that required for pure oscillatory instability. If at that frequency the point on the feedback curve is on the left of the point on the neutron kinetics curves shown on Fig. 2.40, the power P is smaller than the critical power P_c. This is nothing more than the conventional Nyquist criterion, presented graphically in such a way as to keep separate the more accurately known neutron kinetics characteristics and the less accurately known feedback characteristics.

In most reactors it is sufficient to consider the zero frequency point, with coordinates $P \mid X(0) \mid$, $\tau(0)$, to ascertain that the reactor will be stable, if some assumptions are

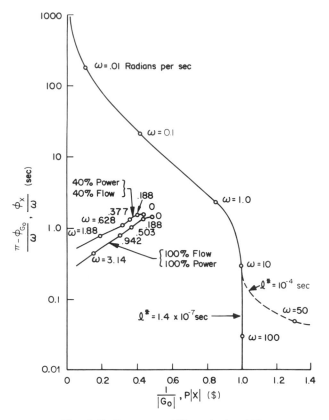

Fig. 2.40. Reactor stability criterion (*10*).

satisfied regarding the frequency dependence of the feedback. These assumptions are:

$$| X(i\omega) | < | X(0) |$$
$$\tau_b(i\omega) < \tau_b(0)$$

(2.53)

for all ω, where $\tau_b(0)$ is defined as $\lim_{\omega \to 0} \tau_b(i\omega)$. If these assumptions hold, a reactor is clearly stable if the zero frequency point $P | X(0) |$, $\tau_b(0)$ is on the left of the neutron kinetics curve shown on Fig. 2.40.

The assumptions are certainly valid for most reactors. If the feedback followed exactly a transport lag model, one would have

$$X(i\omega) = X(0) \, e^{-i\omega\tau_b(0)}$$

(2.54)

so that $| X(i\omega) | = | X(0) |$ and $\tau_b(i\omega) = \tau_b(0)$. This is the limiting case of the assumptions of Eq. (2.53) and the external feedback curve on Fig. 2.39 would be condensed in a point.

Actually the thermal reactivity feedback in solid fuel reactors results from a combination of heat transport and head conduction so that the amplitude $| X(i\omega) |$ of the feedback is attenuated with increasing frequency. Also, as a result of heat conduction, the quantity $\tau_b(i\omega)$ decreases with increasing frequency. For example, in the simple first order heat conduction model, one has

$$X(i\omega) = X(0)/[1 + i\omega\tau_b(0)] \qquad (2.55)$$

so that

$$|X(i\omega)| = |X(0)| \cos \phi_X(i\omega) < |X(0)|$$

and

$$\tau_b(i\omega) = \phi_X(i\omega)/\omega = \tan^{-1}[\omega\tau_b(0)]/\omega < \tau_b(0)$$

The assumptions of Eq. (2.53) possibly would not be valid if the total negative power coefficient consisted of a large positive prompt component and of a larger negative delayed component, as was the case in EBR-I, Mark II, or if the feedback was not of a purely thermal nature but was amplified by a mechanical or hydrodynamic resonance at some frequency. Note that, if the feedback is linear with power, $P|X(0)|$ is the so-called power override, or reactivity, required to bring the reactor from zero power to power P at constant inlet coolant temperature.

For the Fermi reactor the calculated values of $P|X(0)|$ and $\tau_b(0)$ at full power and full flow are, respectively, 49¢ and 1.4 sec. As shown on Fig. 2.40, this point is well on the stable side of the curve. The safety margin is seen to be large, from 49 to 92¢ (the total available excess reactivity) for $P|X(0)|$ and from 1.45 to 14 sec for $\tau_b(0)$. The assumptions of Eq. (2.53) mean that, for increasing values of ω; the point would move toward the left and the bottom of the graph, well away from the curve as shown. Even if the assumptions of Eq. (2.53) were not valid, they would have to be in error by a large amount if the point were to reach the curve, because of the large safety margin. With the feedback mechanism of the Fermi reactor, i.e., where there is no net positive component in the power coefficient, the assumptions of Eq. (2.53) are certainly valid, and the simple stability criterion is just as good as that which would be obtained by a detailed calculation of $X(i\omega)$ based on the same feedback mechanism. The conclusions are only as good as the physical assumptions, but, for given assumptions, the simple method is more conclusive because its simplicity leaves no room for errors.

As indicated by Storrer (12) the behavior of $P|X(i\omega)|$ and of $\tau_b(i\omega)$ can present some anomalies at very low frequencies if some component of the power coefficient has a very large time constant so that it comes into play only at very low frequencies. The hold-down plate expansion coefficient, with a time constant of the order of minutes, is such a component. One can avoid the anomaly by taking for $P|X(0)|$ and $\tau_b(0)$ the fictitious limit which is reached when the frequency is decreased to f_1. As was said previously, this frequency is low enough for all the other components of the power coefficients to have reached their steady state limits, while it is too high for the hold-down plate expansion coefficient to come into play. The values quoted above for the Fermi reactor are those fictitious limits, rather than the time zero frequency limits. The external power coefficient is also a component with a very long time constant. It is irrelevant in this section in which only the internal power coefficient X needs to be considered, since $f > f_1$. However, one should also investigate the stability at other than full flow.

Storrer (12) demonstrates that $\tau_b(0)$ increases somewhat with decreasing flow, while $P|X(0)|$ decreases somewhat if P is reduced in the same proportion as the flow in order to keep the same coolant temperature rise. For instance, at 40% flow, calculations show that $\tau_b(0) = 1.6$ sec.

For many reactors now in operation, both fast and thermal, the point representing the asymptotic feedback characteristics is on the left of the curve shown on Fig. 2.40 so that these reactors satisfy the stability criterion presented here. In boiling water reactors, the power override, which is a good measure of the magnitude of $P \mid X(0) \mid$ and which consists mostly of the reactivity compensated by the voids, can attain many dollars. The zero frequency point is then well on the right curve and a detailed analysis of the frequency dependence of $X(i\omega)$ is required to guarantee stability.

If plutonium or ^{233}U were substituted for ^{235}U in the reactor, and if the power coefficient were about the same in terms of absolute units of reactivity, the numerical value of this power coefficient and the abscissa of the points representing the feedback characteristics of Fig. 2.40 will be multiplied by about a factor of two if the dollar is used as a unit of reactivity. Since, as previously stated, the neutron kinetics curve of Fig. 2.40 will remain approximately unchanged, this substitution would reduce the critical power level for instability by about a factor of two. The plutonium build-up in the core of the Fermi reactor is so small in comparison to the ^{235}U content that no detectable change in the dynamic characteristics will occur.

Stability analysis in the lower frequency range. In the frequency range where $f < f_2$, $X_t = X(0) + \{E_1(0)E_2E_3(0)/[1 - E_2E_4(0)]\}$, and the power P_c, at which the denominator of Eq. (2.43) becomes zero should be determined,

$$1 - P_c G_0 X_t = 0 \tag{2.56}$$

Noting that $E_4(0) = 1$ and defining α as

$$\alpha = E_1(0)E_3(0)/2X(0) \tag{2.57}$$

one obtains

$$X_t = X(0)\{1 + [2\alpha E_2/(1 - E_2)]\} \tag{2.58}$$

As shown below, the following relations

$$0 < \alpha < 1 \tag{2.59}$$

$$\mid E_2 \mid < 1 \tag{2.60}$$

are valid for any frequency and, with these two conditions, Eq. (2.56) can never be satisfied for any power or frequency. Hence, system stability is assured at any power level.

The relation $0 < \alpha < 1$ can be obtained in the following way. If the steady state reactivity feedback caused by a unit power change $X(0)$ is identical to the reactivity feedback caused by an isothermal temperature increase of the whole reactor equal to the increase of the average coolant temperature, one would have

$$X(0) = E_1(0)E_3(0)/2 \tag{2.61}$$

since $E_1(0)/2$ represents the increase of the average coolant temperature per unit power increase and $E_3(0)$ is the isothermal temperature coefficient of reactivity. By combining Eqs. (2.57) and (2.61), one can see that $\alpha = 1$. Note that the relation (2.61) would approximately hold for a symmetrical homogeneous reactor. Since, in a symmetrical heterogeneous reactor the fuel temperature at power is higher than that of the coolant, $X(0)$ is certainly larger than in the homogeneous case and α must be less than 1. $\alpha > 0$ means simply that the power coefficient and the isothermal temperature coefficient have the same sign.

In the Fermi reactor the calculated values are:

$$E_1(0) = 1.25°F/MW \text{ at full flow}$$
$$E_3(0) = 0.291¢/°F$$
$$X(0) = 0.2458¢/MW \text{ at full flow}$$

and thus $\alpha = (1.25 \times 0.291)/(2 \times 0.2458) = 0.74$ at full flow. At reduced flow the value of α is somewhat higher, but still smaller than unity, for the reason given above. E_2 is the transfer function giving the change in reactor inlet coolant temperature resulting from the transmission around the coolant loops of a change in the reactor outlet coolant temperature. The gain E_2 of this transfer function is certainly smaller than unity, since a passive thermal system can never act as an amplifier.

From Eqs. (2.56) and (2.58), it is clear that, since the phase of the zero power transfer function G_0 never exceeds 90°, and since $X(0)$ is a negative number, Eq. (2.56) can only be satisfied if the phase of $1 + [2\alpha E_2/(1 - E_2)]$ is at least 90°. That this is impossible, and that therefore instability is impossible when $f < f_2$, is demonstrated in the following paragraph.

$1 + [(2E_2\alpha)/(1 - E_2)]$ can be written as $[1 + (2\alpha - 1)E_2]/(1 - E_2)$. For the case where α has its maximum value of unity, this expression reduces to $(1 + E_2)/(1 - E_2)$. From a simple geometric construction it can be seen that the phase of $(1 + E_2)/(1 - E_2)$ can never exceed 90° when $|E_2| < 1$, whatever the phase of E_2 itself is. In Fig. 2.41, the geometric constructions of $1 + E_2$ and $1 - E_2$ are given for an arbitrary phase angle for E_2. The particular phase angle that was used is not important because, as will be seen

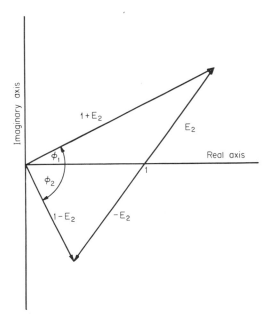

Fig. 2.41. Geometric construction (*10*) for determination of the phase angle of $(1 + E_2)/(1 - E_2)$.

later, the results are independent of the phase angle of E_2. The phase angles of $1 + E_2$ and $1 - E_2$ are indicated as ϕ_1 and ϕ_2, respectively. The former is measured in a counter-clockwise direction and is considered to have a positive value; the latter is measured in a clockwise direction and is considered to have a negative value. The phase angle of $(1 + E_2)/(1 - E_2)$ is $(\phi_1 - \phi_2)$ by the rules of complex arithmetic. Consider now the case when $E_2 = 1$. For this case ϕ_1 and ϕ_2 will be at their respective maxima in an absolute sense, since, by physical reasoning, $|E_2|$ can never be greater than unity. Therefore, $(\phi_1 - \phi_2)$ will also have a maximum when $|E_2| = 1$. Now, when $|E_2| = 1$, the line formed by the vectors E_2 and $-E_2$ can be considered to be the diameter of a circle and the vector from the origin to unity a radius. Then the above-mentioned diameter and the vectors $1 + E_2$ and $1 - E_2$ become the sides of a triangle inscribed in a semi-circle. By geometric reasoning this triangle is a right triangle and the angle $(\phi_1 - \phi_2)$ is a right angle. Thus, for this case, $(\phi_1 - \phi_2) = 90°$. By inspection it is obvious that when α is less than unity and $|E_2|$ is less than unity the included angle at the origin between the adjacent sides of the resulting triangle will be less than 90°. This conclusion is valid for any value of the phase angle of E_2. Thus, there is no possibility of instability when $f < f_2$.

2.6 External Influences

The previous sections have outlined general disturbances to the reactor core. These disturbances may be ameliorated or aggravated by the control system or by human action.

2.6.1 SURVEY OF TYPES

It is worth surveying possible initiators for the general disturbances so far considered, in particular, external initiators.

(a) Flow perturbations might arise from

   (1) loss of electrical power to the plant or plant components resulting in a loss of pumping power;
   (2) pump mechanical failures;
   (3) control malfunction.

(b) Reactivity perturbations might arise from

   (1) the introduction of bubbles into the system;
   (2) seismic deformations through structural movement;
   (3) control malfunction;
   (4) refueling accident.

(c) Thermal perturbations might arise from

   (1) loss of feedwater supply;

(2) feedheater or turbine malfunctions resulting in a loss of heat removal capability;

(3) control malfunction in secondary and tertiary loops.

In each case a control malfunction is a possible cause of an accident, and as a result a controller error is treated in any safety assessment. The actual course of events depends critically on the mode of control used in any particular plant.

2.6.2 CONTROL MODES

The control philosophy is shaped by certain restrictions which are set by plant material considerations and by the plant characteristics themselves.

The main requirements are: (a) a constant turbine stop-valve temperature, and (b) minimized temperature gradients in the vessel and in primary components. Whether or not these requirements are achievable depends on the time constants throughout the circuits of the system.

In a steam-cooled system which is a partly indirect cycle by virtue of the reheat cycle, load-following is possible if the feed supply-valve is administered correctly while the turbine stop-valve is varied. If the time constants and temperature feedback coefficients are suitable then this could be the control philosophy.

In a liquid-metal-cooled fast breeder the long delays in the intermediate circuit combined with positive void coefficients rule out load-following. A load-setting procedure is therefore a necessary control mode. This might take the following form: set the flow in the primary and secondary circuits; adjust the feed circuits; and adjust the reactivity in the core through control absorbers to meet the main control requirements listed above. The main problems are the difficulty in getting temperature signals from the core and the need to optimize control rod movement for small temperature changes.

Thus in the LMFBR control system, malfunctions might give rise to disturbances in primary, secondary, or feed flows, in reactivity, or possibly in a combination of all four.

In a given design a fault-tree analysis of the protective logic and the control system will be needed to decide on possible combinations of control operations that could give rise to adverse core effects. A faulty reduction of flows in the heat transfer loops following a signal of increased outlet temperatures would be an example of a gross control malfunction. The analysis of such possible malfunctions will need a combination of the analytical techniques discussed in Chapter 1.

One other possible control mode incident aggravation is worth mention. Noted above was the need to optimize control rod movement for small temperature changes. This optimization sometimes means the movement of a number of control rods in a staggered mode that makes it difficult to know what reactivity state the system is in at any given time. This ignorance of the reactivity state of the system hampered the diagnosis of the Fermi incident (see Section 4.6) which occurred during start-up.

Each control system must be treated as a special case for analysis.

2.6.3 SCRAM PROTECTION

Scramming a reactor causes thermal shocks to the system and there is need to minimize the number of unnecessary fast scrams. Section 3.4 details the place of scrams in safety analysis and the possibility of using adverse reactor signals for warnings and power setbacks rather than scrams. Here the discussion is concerned with scram participation in the dynamic analysis.

2.6.3.1 *Scram Delays in Instrumentation*

Each type of scram has a combined delay which depends on the delays in obtaining signals from the system and delays in acting upon these signals.

(a) System physical *transport delays* are usually small except for the transmission of temperature in the reactor, plenums, and heat exchangers. The transport of fission products from failed fuel to fission product and delayed neutron detectors elsewhere in the system also incurs a delay.

(b) *Instrument delays* vary with the instruments in question.

(1) Reactivity instrumentation, flux, and period meters incur a delay due to noise filtering; the latter incur the more serious delays as they need more filtering.

(2) Thermocouples have varying delays depending on whether the couple is sheathed or not. A $\frac{1}{8}$-in. ungrounded thin sheath thermocouple in a LMFBR might have a 1000 msec delay with a further 10 msec for its amplifier.

(3) Flux/flow ratio measurement delays depend on the fact that separate measurements are divided. Electronic dividers may have no delay, whereas electromechanical servos might add 100 msec and thermal dividers could add 500 msec.

(4) Tachometer measurements are rapid; a dc tachometer filter delay would be about 5–10 msec whereas an ac induction tachometer can have filter delays from 5 to 50 msec.

(5) Flowmeters can have delays of 10 msec or less.

(6) Pressure meters incur a delay through the diaphragm and isolation device which may be between 250 and 500 msec, depending on the capillary tubing used. There may be an added 100 msec sensor delay.

(7) Under- or overvoltage relays would have to be protected from spurious operation due to surges and false alarms by time lags which could amount to 500 msec.

(c) *Logic delays* following the reception of the signal would typically be 20 msec.

(d) Once the signal has been received and analyzed and dispatched to the shut-down mechanism, a further *rod release delay* including the scram breaker dropout would add a further 120 msec.

2.6.3.2 *Total Scram Delays*

From these various component delays we obtain the total typical delays in a LMFBR system shown in the accompanying tabulation.

Signal	Instrument	Delay (msec)
Reactivity detection	Flux meter	300
	Period meter	330
	Flux/flow ratio	700
	Outlet thermocouple	2300
Loss of flow detection	Flowmeter	300
	Shaft tachometer	300
	Pipe pressure	500
	Vessel level	500
	Pump power relays	1300
	Outlet thermocouple	2300
	Flux/flow ratio	700
	Bypass flowmeter	300
Loss of heat removal	IHX outlet primary temp.	2300
	Secondary flowmeter	500

The temperature signals are noticeably slower and have additional physical delays which are not included in the tabulated figures. These transport delays range from 500 msec in the IHX measurement to 1000 msec at the outlet plenum or even longer at low flow levels.

Scrams using such combined delays as these are used in the safety evaluation representations.

2.6.3.3 *Component Defects*

Most component defects, be it in material, design, or fabrication, can be considered as external disturbances to the system which are built in but which may appear at any time, especially when that material or component is operating under adverse off-normal accident conditions. Such component defects take three major forms as we shall note in the coming chapters.

(a) *Fuel cladding defects.* Under typical quality control procedures and methods it is only possible for a one mil defect in 15 mil cladding to avoid detection. This size of defect has no significant effect on the strength of the cladding (*13*).

(b) *Bad design.* Section 4.6 will detail examples of bad design of components, procedures, and operations. Fault trees attempt to take into account all bad design possibilities and such errors are included in the safety evaluation. However, bad design, being a human failing, is always a present possibility against which safeguards are provided.

(c) *Lack of quality assurance over materials.* Examples have occurred of materials which, through lack of quality control at the original suppliers, were not produced according to specification. An example is the inclusion of one carbon steel tube among 3600 stainless steel tubes for the Enrico Fermi reactor steam generators. Such a tube could have failed and resulted in further failures if not excluded from the component. The safety evaluation will include the consequences of material being other than as specified in strengths, heat-transfer coefficients, erosion and corrosion resistances, etc.

REFERENCES

1. J. C. R. Kelly, A. Biancheria, C. A. Anderson, R. G. Hobson, and E. F. Beckett, The Westinghouse 1000 MWe—Follow-on Study, WARD-5703. Westinghouse Advanced Reactors Division, Waltz Mill, Pennsylvania, November 1968.
2. C. P. Gratton, E. G. Bevan, J. Graham, R. Hobday, and A. T. Hooper, A gas-cooled fast reactor using coated particle fuel. *J. Brit. Nucl. Energy Soc.* **7** (3), 233 (1968).
3. R. A. Webb and D. C. Schluderberg, Steam cooling for fast reactors. *Proc. Conf. Safety, Fuels, and Core Design in Large Fast Power Reactors, Argonne Nat. Lab. October 1965*, ANL-7120. Argonne Nat. Lab., Argonne, Illinois, 1965.
4. P. Fortescue, R. Shanstrom, J. Broido, J. M. Stein, A. Baxter, and H. Fenech, Safety characteristics of large gas-cooled fast power reactors. *Proc. Conf. Safety, Fuels, and Core Design in Large Fast Power Reactors, Argonne Nat. Lab., October 1965*, ANL-7120. Argonne Nat. Lab., Argonne, Illinois, 1965.

5. J. Zoubek, Private communication, 1970.
6. S. O. Arneson, Design features and testing capabilities of the Fast Flux Test Facility. *Int. Conf. Fast Reactor Irradiation Testing, Thurso, Caithness, April 1969*, Paper 8/1.
7. L. E. Weaver, "System Analysis of Nuclear Reactor Dynamics." Rowman and Littlefield, New York, 1963.
8. J. LaSalle and S. Lefschetz, "Stability by Liapunov's Direct Method." Academic Press, New York, 1961.
9. E. P. Gyftopoulos, Lagrange stability by Liapunov's direct method. *Reactor Kinet. Control* 2, 227 (1964); also as USAEC TID 7662. U.S. At. Energy Comm., April 1964.
10. T. J. Thompson, Accidents and destructive tests. *In* "The Technology of Nuclear Reactor Safety" (T. J. Thompson and J. G. Beckerley, eds.), p. 625. M. I. T. Press, Cambridge, Massachusetts, 1964.
11. Enrico Fermi atomic power plant: Technical information and hazards summary report. USAEC Rept. NP-11526, Pt. B, Vol. 7. U.S. At. Energy Comm., Washington, D.C., 1961.
12. F. Storrer, Analysis of the power feedback relations in fast reactors. *Proc. Conf. Transfer Function Measurements and Reactor Stability Analy., May 1960*, ANL-6205, p. 251, Chicago, 1960.
13. K. C. Thomas, Private communication, Westinghouse, 1969.

CHAPTER 3

SAFETY CRITERIA

3.1 Failure Criteria

The first two chapters have emphasized methods of safety evaluation and possible disturbances to the system. The next chapters go through a design process by setting design safety criteria, by reviewing the particular problems of a sodium-cooled design, by containing the system, and finally by licensing the plant.

Before reviewing the safety criteria that are set to define the design of a safe plant and detailing safety features that might be included, it is necessary to consider what failure means. In the previous chapter the system was evaluated to determine its response to a variety of disturbances; in particular to assess whether failure was possible. What is meant by failure?

3.1.1 FUEL FAILURE CRITERIA

A typical fuel pin in a fast reactor could be a ceramic fuel bonded within a metal cladding. Several varieties are possible, the main ones being helium-bonded UO_2 (enriched with PuO_2) in stainless-steel cladding; and sodium-bonded UC pellet in stainless-steel cladding. The oxide and carbide can be fabricated in different ways (pelleted or vibro-compacted) and the cladding can be fabricated in different ways (solution-treated or cold-worked). Many other varieties are possible even in different configurations ranging from cermet fuels or silicon carbide-coated fuel particles (*1a*). Each of these fuel elements has a different mode of failure depending on the circumstances and so different failure limits are required for each fuel element for each disturbance.

3.1.1.1 *Critical Heat Flux Criterion*

A reduced flow rate results in reduced heat removal and eventually damage to the fuel pin. If the pin is sodium-bonded then this damage will first appear as bond vaporization with consequent insulation of the fuel. Thus failure of the fuel pin might be considered to start when the fuel surface heat flux exceeds a critical value for sodium vaporization. This is rather similar to pool boiling and the critical heat flux (CHF) is given by Eq. (3.1).

$$\text{CHF} = 534{,}000\, p^{0.17} \quad \text{Btu/hr-ft}^2 \tag{3.1}$$

where p is pressure (psia) (*1b*).

Such a failure criterion corresponds to PWR burn-out conditions. The criterion can be modified by local conditions in any particular design and may possibly be correlated to fuel surface temperature for a particular bond pressure. In one design in which it was calculated that when the fuel surface temperature exceeded 1900°F, the critical heat flux was attained, this surface temperature could itself be used as a failure criterion.

Bond boiling would effectively insulate the hot fuel until fuel melting had occurred. The molten fuel then might come into contact with the cladding, so that although vaporization of the bond might not burst the cladding, molten fuel in contact with the cladding or fuel vaporization would.

This failure criterion does not however apply to gas-bonded oxide fuel pins and other failure criteria are required.

3.1.1.2 *Cladding Rupture Mechanisms*

In a gas-bonded fuel pin, failure might be defined as being coincident with cladding rupture, which, if the pin were originally unvented, would allow the release of fission gases.

The rupture will depend on conditions within the fuel pin that depend in turn on the power history of the pin and its burn-up. The structural changes may be summarized as follows:

a. *At start-of-life.* Figure 3.1 shows a cross section of a gas-bonded, stainless-steel-clad oxide pellet with a fuel density of from 85 to 95% theoretical density. The design includes a 5 mil cold gap between the fuel and the cladding.

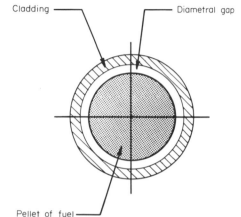

Cladding   Diametral gap

Fig. 3.1. Structure of fuel, cold and at the start-of-life.

Pellet of fuel

b. *At low burn-up.* Fuel above the sintering temperature (3272°F) sinters and lenticular voids migrate up the temperature gradient and a central void is formed. The fuel above the sintering temperature attains the theoretical density and columnar grains mark the migration routes of the lenticular voids. Fuel at less than the sintering temperature remains at its original density. The void migration rates fall off rapidly over a small radius increment.

c. *At higher burn-up.* Continued irradiation (to about 25,000 MWD/tonne) causes fuel swelling and fission gases are generated. The rate of fuel swelling is higher than that of the cladding and so the fuel meets the cladding and then there will be a fuel–cladding contact in the higher rated regions. Fission gases are released from the fuel and they diffuse to the fuel pin gas plenum (Fig. 3.2).

d. *Later in life.* At about 70,000 MWD/tonne the central void starts to decrease and could close eventually. The cladding at the end-of-life is weaker due to thermal cycling, erosion, and corrosion. Reductions in cladding tensile strength have been reported of between 15 and 30% over 90,000 hr irradiation (*2a*), and the strain on the cladding might be as much

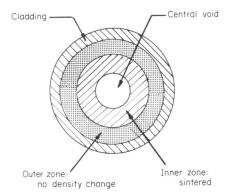

Cladding —

Central void

Outer zone:
no density change

Inner zone:
sintered

Fig. 3.2. Structure of fuel at end-of-life.

as $\frac{1}{2}\%$ (*2b*). At this stage the fuel pin is most sensitive to adverse conditions as in addition, the pressure of fission gases within the fuel pin plenum may be between 800 and 3500 psia depending on the size of plenum (*2a*).

It is assumed that the amount of fission gas released from the fuel is a function of the temperature of the fuel (*3a, b*): 100% is released at temperatures over 3272°F; 50% is released at temperatures between 2912°F and 3272°F; and 4% is released at temperatures less than 2912°F. Thus the total gas released can be calculated by integrating over the fuel temperature distribution. Values of about 65–70% can be expected. During a transient when the temperatures increase little additional gas will emerge.

In addition to this normal operational data, the TREAT facility has provided additional information on abnormal conditions in a transient (*4*):

(a) Cladding deformations of greater than 1% result in gross cracking on the inner cladding surface due to heavy grain precipitation.

(b) No foaming of irradiated oxide fuel occurs even when 70–80 vol % of the fuel is melted.

(c) Agreement with the calculated temperatures is fairly good. This indicates that the fuel condition can be successfully modeled.

(d) In irradiated fuel, the failure mechanism appears to be cladding melting due to contact with molten fuel.

With this information the following failure criteria may be derived.

3.1.1.3 *Reactivity Accident*

When the fuel center temperature attains 6500°F the vapor pressure starts to rise rapidly from 10 to 100 atm at 7250°F. Under these conditions the high pressures could force molten fuel through cracks into contact with cladding and cause cladding failure.

On the other hand, if there is a large amount of fuel melting the excessive fission gas release will cause excessive pressures on the cladding. The amount of molten fuel judged to be excessive depends, of course, on the condition of the cladding and the burn-up of the fuel.

Thus at start-, middle- and end-of-life, a fuel temperature of 6500°F may be judged to portend failure. However, 60 mil of fuel melting (25% areal extent) represents failure at the start-of-life while 20 mil may represent a similar failure at the end-of-life due to the weakening of the cladding and the rise of fission gas pressures.

The position of the failure depends on the pattern of fuel melting (5).

The modeling and the analysis is borne out by experimental results of failures and by radiographs of fuel structure and cladding cracking at various burn-up levels. This latter information is unfortunately only obtained when the fuel has cooled off, of course.

3.1.1.4 *Loss of Cooling Accident*

A simpler criterion would now be more valid, either one simply related to the cladding melting temperature in the range of 2500°F, or a strain due to thermal stresses (say 0.5 or 1% strain) at 1700 to 1800°F, or the boiling point of the coolant sodium (1632°F at atmospheric pressure). The reasoning for the latter (5) is that although sodium boiling does not infer cladding rupture, conditions are probably such that cladding failure is not far away.

3.1.1.5 *Accident Severity Levels*

It is now possible to define accident severity levels for gas-bonded oxide fuel in stainless-steel cladding. Table 3.1 shows typical severity levels for

TABLE 3.1

CORE DAMAGE SEVERITY LEVEL CLASSIFICATION

Damage severity level classification	Cross-sectional area of molten fuel in the pin (%)		Cladding temperature maximum design (°F)
	Start-of-life	End-of-life	
Gross	75	40	1600
Moderate	25	3	1400
No damage	0	0	1300

start- and end-of-life conditions. These may be used in a classification of core accidents (Section 3.1.3). The actual values will depend critically on the fuel pin design.

3.1.2 STEAM GENERATOR FAILURE

Other components of the plant can fail; one of the more probable being the steam generator in which a sodium–water interaction becomes a possibility in the LMFBR system.

TABLE 3.2

STEAM GENERATOR FAILURES

Failure characteristic	Initiation	Primary or secondary failure
(1) Mode	Erosion	P
	Corrosion	P
	Rubbing (fretting)	P
	Vibratory fatigue	P
	Thermal fatigue	P
	Creep to rupture	P
	Thermal buckling	P
	Failure associated with excess creep strain	P
	Cracks due to material	P
	Cracks due to fabrication	P
	Due to sodium–water reaction: tube wastage	S
	Vessel rupture	S
	Tubes torn away	S
(2) Location	Tube to tube	
	Tube to tube support	
	Tube	
	Tube to tubesheet	
	Shell	
	Shell to shell nozzle, sodium side	
	Shell to tubesheet nozzle, steam side	
	Tube support or baffle	
(3) Cause	Design	
	Material specifications	
	Manufacture	
	Poor fabrication and welding	

Table 3.2 shows the modes, the locations, and the causes of failures experienced to date, some of the failures being primary and some secondary as noted. Such a survey of nuclear facilities gives some indications of practical experience. Out of 16 facilities, only 6 reported failures, and all of these were due to faults that could have been caught by adequate quality assurance and inspection techniques, rather than being due to behavior outside normal operating conditions. The locations of the failures were mainly in the tube sheet or at the tube-to-tube sheet joint.

From this information a failure rate can be calculated for sodium and nonsodium steam generators: (a) sodium operation, 21 failures per 10^9 tube-hr; and (b) non-sodium operation, 29 failures per 10^9 tube-hr (6). Thus the experience with sodium is not significantly different from the experience in water-to-steam systems. This failure rate would imply a failure about every 3 or 4 years for a 700 MWt plant.

This information can be used to assess availability and maintainability and can perhaps be used to establish some quantitative failure probabilities for fault-tree analysis. However such information cannot be used for establishing the point of steam generator failure during abnormal operation.

Similar failure rates can be established for all other components in a nuclear power plant and this is currently being organized in the US by the Liquid Metals Engineering Center (7). The center collects all failure data available for liquid-metal systems and analyzes the incidents for cause and origin. Naturally, statistics are still poor, but Table 3.3 shows some of the results available. Note how misleading the failure rate for diesel generators might be since it is as yet derived from only four failures. The low failure rate for core fuel and breeder elements is worth noting because almost all the data arise from fuel failures in the Enrico Fermi plant.

The accumulation of failure data is a necessary part of safety engineering. At present it should be considered as long-term data which will be available when fault-tree analyses for plant failures become quantitative sometime in the future.

3.1.3 ACCIDENT CLASSIFICATION

Knowing something of the failures which might occur and something of the response of the system to the accidents, it is important to classify the accidents according to:

(a) *Probability*. This is impossible to do exactly at present and engineer-

TABLE 3.3

FAILURES OCCURRING DURING FACILITY OPERATION[a]

System	Component involved	Failure rate (per 10^6 hr)	Number of failures
Chemical	Demineralizers	340	6
Containment	Personnel air locks	290	6
	Hoist units	166	8
	Shielding	110	9
	Vessel internals	200	8
	Vessels and tanks (sodium)	120	7
Electrical	Diesel electrical generators	1640	4
	Heaters	125	16
	Motors	120	18
	Power switch gear	105	2
	Circuit breakers	310	14
	Power relays	110	7
	Transformers	275	4
	Turbine generators	80	10
Energy conversion	Furnace equipment	105	6
	Pneumatic motors	220	1
	Drive shafts	125	60
	Turbine generators	101	3
	Valve operators	200	5
Fuel handling	Fuel and breeder elements	49	19
	Fuel handling equipment	75	3
Heat transfer	Blowers and fans	140	17
	Cold traps	220	8
	Hot traps	410	1
	Coolers	120	1
	Desuperheaters	260	9
	Feedwater heaters	180	3
	Filters and strainers	120	2
	IHX	32	2
	Sodium piping	260	7
	Piping supports	104	5
	Pumps and supports	325	80
	Rupture disks	310	2
	Steam generators	125	8
	Traps for sodium	230	22
	Sodium valves	120	105

TABLE 3.3 (*continued*)

System	Component involved	Failure rate (per 10^6 hr)	Number of failures
Heat transfer	Feedwater valves	305	50
	Steam valves	280	40
	Air driers	210	2
Instrumentation	Sodium level annunciators	210	2
	Compressors	200	42
	Safety rods	187	3
	Electrical controllers	210	22
	Electronic indicators	100	17
	Mechanical indicators	70	4
	Pneumatic indicators	280	4
	Sight glass indicators	295	55
	Steam indicators	285	4
	Instrumentation relays	275	2
	Neutron sources	126	1
	Temperature sensors	149	3
	Pressure sensors	181	11
	Conductivity sensors	89	3
	Neutron monitor sensors	205	4
	Fission gas sensors	60	2

[a] See Budney (6).

ing judgment is used. As failure statistics improve, then it may be done for some of the plant components.

(b) *Severity.* This is a calculated severity with experimental confirmation where possible.

Light water reactors (LWR) also use accident classifications that range from those accidents and occurrences which are expected frequently (minor perturbations in reactivity or single failures of control elements) to those which are never expected to occur but which are nevertheless chosen as a design basis because their consequences are so severe (major pipe break, ejection of a control rod).

An accident classification may tie the probability and severity of the consequences together, without reference to criteria for the design of the plant. A more logical classification is proposed which in general may be applied to any nuclear system.

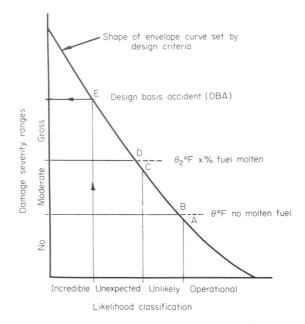

Fig. 3.3. Accident classification in terms of damage severity and likelihood.

3.1.3.1 *Proposed Core Damage Severity Ranges*[†]

Three damage ranges are specified without reference to the accidents which could cause them and without reference to any probabilities for those accidents.

(a) *No damage range.* Within this range incidents give rise to slightly above normal temperatures and stresses but no physical damage occurs and there is no significant reduction in attainable burn-up. There is no fuel melting and the peak cladding temperature is less than $\theta_1°$F.

(b) *Moderate damage range.* Within this range some (up to $x\%$ cross-sectional area) fuel melting occurs but there is no loss of fuel pin integrity (maximum peak cladding temperature of $\theta_2°$F). No fission gas is released but there is a reduction in the attainable burn-up and possibly in transient operational capability.

(c) *Gross damage range.* Within this range substantial fuel melting occurs (greater than $x\%$ of the cross section). Cladding failure together with a release of fission products and fuel material may occur. At the upper end of this damage range, this failure may be violent and combined with a loss

[†] See **Fig. 3.3**.

of core, and, possibly, an explosive disruption of the core. The upper limit of damage is defined as that arising from the worst credible accident.

3.1.3.2 *Likelihood Classification*[†]

Now one can define a fault classification which is based on a judgment of probabilities within the following ranges.

(a) *Operational occurrences.* These are off-normal conditions which individually may be expected to occur during the plant lifetime. (The plant design and protective system is such that the effects of operational faults will be limited to lie within the no damage range but this criterion is not part of the definition of an operational fault.)

(b) *Unlikely faults.* These are off-normal conditions which individually are not expected to occur during the plant lifetime but which, when integrated over all components of the reactor, one such fault may be expected to occur once during the plant lifetime. It is probable that such an unlikely fault would simply lead to the necessary repair or replacement of the faulty component or system. (Plant and protective system design criteria will limit the consequences of unlikely faults to the no damage or moderate damage ranges but these criteria are not part of the definition of an unlikely fault.)

(c) *Unexpected faults.* These are off-normal conditions arising from a single very low probability failure which are expected never to occur during the plant lifetime but which are nevertheless mechanistically possible.

By virtue of the design of the plant and its safeguards some of these unexpected faults will represent extreme cases of failure which are identified as being possible but of extremely low probability. The worst of these credible[†] accident situations is chosen as the plant design basis accident. Its consequences form the upper limit of damage in the gross damage range. Plant design criteria seek to reduce this damage level.

(d) *Incredible faults.* These are off-normal conditions which arise from two or more very low probability failures or which are postulated to occur irrespective of the fact that no credible[‡] initiator can be envisaged. These accidents are presumed to be of insignificant probability and therefore would not be used as design basis faults.

[†] See Fig. 3.3.
[‡] Credible is defined as being mechanistically founded, arising from a real, even though low probability, fault initiator.

In the severity classification, θ_1, θ_2, and x will depend on the fuel design chosen but they might be expected to be of the order of:

$\theta_1 = 1300-1400°F$,

$\theta_2 = 1500-1600°F$,

$x = 20-40\%$ (see Table 3.1).

3.1.3.3 *Design Criteria*

Having defined severity ranges and an accident likelihood classification, the scales of a graph have been established onto which accidents and their consequences can in principle be plotted. There will be a whole galaxy of points but because the plant has been well designed, the points will tend to fall towards the origin. The expected faults will contribute only minor consequences whereas severe consequences could come from only the very low probability accidents. This good design is supported by design criteria which set the shape of the curve which envelops all the plotted occurrences (Fig. 3.3). The following criteria will define the curve shown in the figure.

(a) No operational occurrence in the system shall result in consequences worse than the no damage range. Thus point A must lie below point B (θ_1).

(b) No unlikely fault in the system shall result in consequences worse than those in the moderate damage range. Thus point C must lie below point D (θ_2, x).

(c) The unexpected fault with the worst consequences shall be designated the design basis accident for the system. Thus point E sets the upper limit of the gross damage range.

(d) Incredible faults are not considered in the safety evaluation beyond showing them to be incredible according to the definition. The design does not accept their existence.

3.1.3.4 *Event Classification*

It now remains to put accidents to the plant in their respective probability ranges. Table 3.4 gives a prospective listing of accidents, but this list is by no means final for any given plant, each system having to be considered on its own merits. Such a list is prepared on the basis of engineering judgment, although in the future it is expected that a qualitative classification will be defensible.

TABLE 3.4

CLASSIFICATION OF EVENTS BY LIKELIHOOD

Likelihood class	Event
Operational occurrences	Reactivity control withdrawal error at power or start-up
	Random fuel failures
	Human error
	Flow control error
	Loss of one pump flow due to loss of electrical supply
	Small seismic shock
	Feedwater heating supply failure
	Turbine stop-valve closure
	Loss of cover gas pressure
	Steam generator tube failure
	Offsite power failure
	Loss of onsite auxilliary power
Unlikely faults	Failure of control rod hold-down mechanism
	Operating basis earthquake
	Pump mechanical failure
	Small pipe leaks
	Large bubbles reaching the core
	Loss of primary scram following an operational occurrence
	Independent active component failure following an operational occurence (see Section 3.3.4)
Unexpected faults	Large system ruptures
	Large sodium fire
	Failure of core support allowing core movement
	Design basis earthquake
	Very large bubbles reaching the core
	Passive component failure following an operational occurrence (see Section 3.3.4)
	Loss of scram following an operational or unlikely fault
	Loss of offsite and onsite power supplies
Incredible faults	A refueling accident leading to criticality
	Rod ejection
	Sudden local subassembly blockage

3.1.4 SURVIVAL CRITERIA FOR FUEL PINS

The previous sections have discussed failure criteria from the point of view of anticipated failure mechanisms arising from faults. We also need to know how fuel pins may survive under normal conditions and so cladding survival criteria can be defined in terms of the following.

(a) Once-off conditions due to operational transients which might result in: overstress due to internal loadings; overtemperatures due to high power-to-flow conditions; and defect failure which is exhibited as an aggravation of other modes of failure. These conditions all arise from minor accidents or transient overshoots such as those discussed in Chapter 2.

(b) Cyclic conditions due to repeated minor loadings: stress cycles; temperature cycles due to load changes; external loadings, possibly from flow disturbances; and vibrational fatigue from flow induced vibrations.

(c) Continuous adverse conditions due to normal operation: cladding erosion; cladding corrosion; and internal loadings due to fuel swelling and fission gas pressure.

In these cases the cladding would have a failure because of a reduction in allowable cladding strain before yield. This might, in combination, reduce the failure limits discussed in Section 3.1.1.

3.1.5 PROTECTIVE SYSTEM SETTINGS

To get some perspective of plant and core level limitations, Table 3.5 shows a typical set of trip points for a LMFBR system. These levels also constitute a set of failure criteria, at least as far as continued operation of the plant is concerned. The operator is concerned with remaining inside these values, which necessarily means that he will also not approach the previously discussed fuel pin failure criteria because a considerable margin of safety is provided by the choice of the low trip points.

Other trips may be possible and desirable. However it must be remembered that too many trips constitute a safety hazard because they cause operator frustration in shutting down the reactor when the operator's job is to keep it operating safely. This situation is worse on an experimental system, where the operators are under considerable pressure from experimentalists to maintain constant power conditions.

A power rate-of-change trip (period meter) is sometimes required as it gives a very early indication of things beginning to change. However the device is noisy [being a divider to give $P(dP/dt)^{-1}$] and is not particularly

TABLE 3.5

PROTECTIVE TRIP SETTINGS

Trip signal	Range	Trip level	Redundancy[a]
High nuclear flux	Power range	110%	2/4[b]
	Intermediate range	10%	2/3
	Low level	100 kW	2/3
	Source level	1 W	2/3
Flux-to-flow ratio		1.20	2/4
Low flow		80%	2/4
High core inlet temperature		10% of core temp. rise	2/4
High reactor outlet temperature		Same	2/4
Low reactor vessel level		−1 ft	2/4
Seismic activity		MM V[c]	2/4
Loss of electrical power		Yes	1/2
Containment high pressure and radioactivity		Yes	2/4
Manual		Yes	1/1

[a] Varies considerably from plant to plant. See Table 3.6.
[b] Slash in 2/4 stands for "out of."
[c] Modified Mercalli V.

in favor among control designers. A reactivity meter which derives the reactivity as a function of time from measurements of flux as a function of time is a requirement in modern LMFBR systems.

3.2 General Safety Criteria

3.2.1 AEC CRITERIA

In July 1967, the AEC published a set of 70 design criteria for use by nuclear power plant designers. The AEC had the following objectives in doing this: (a) the provision of a record of the criteria which had already been used by designers to that date; (b) the provision of a standard set of criteria so that all designers could follow the same rules. This would make licensing somewhat easier; and (c) the provision of a safety check list to ensure that all safety considerations would be covered before a license was issued.

The criteria were published as "General Design Criteria for Nuclear Power Plant Construction Permits" (*8*). These criteria were reduced to 58, rewritten and republished in February 1971 (*9*). They are split into six general categories.

	Criteria
I. Overall plant requirements	1–6
II. Protection by multiple fission product barriers	10–19
III. Protection and reactivity control systems	20–29
IV. Fluid systems	30–46
V. Reactor containment	50–57
VI. Fuel and radioactivity control	60–64

The criteria certainly fulfilled the objectives above but they had certain disadvantages for fast reactor designers.

(a) The criteria were too vague, partly because of the state-of-the-art and partly because of a desire not to hamper the designer in his design choices.

(b) They apply to the nuclear reactor of the day: the light water reactor (LWR) system. They do not necessarily apply to fast systems, although many of the criteria are general enough to include all reactor systems.

(c) They were almost law and therefore difficult to change. Fast reactor designers have undoubtedly obtained relevant criteria of their own, but no AEC criteria specifically for fast reactors have yet been prepared. However the American Nuclear Society is moving to prepare such criteria through one of its standards subcommittees.

3.2.2 RADIOACTIVITY RELEASE CRITERIA

In addition there are two other sets of regulations published in the Federal Register. One applies under normal operations (10 CFR 20) and the other set is for extremely unlikely accident conditions (10 CFR 100). They detail the allowable releases in these situations. These will be discussed further in Sections 5.1 and 5.2. In the 1967 criteria, Criterion 70 was the overriding criterion because it simply stated that the design must be such as to comply with these two radioactivity release regulations, 10 CFR 20 and 10 CFR 100. However, in the 1971 criteria, Criterion 60 ensures that the design will "maintain suitable control" over radioactive releases during expected occurrences while Criteria 16 and 50 ensure that containment will be provided with a suitable low leakage rate during accident conditions.

3.2.3 FAST REACTOR CRITERIA

The following differences exist which make some of the AEC general criteria inapplicable to the liquid-metal-cooled fast reactor systems.

(a) The LMFBR is a low-pressure system whereas the LWRs are high-pressure systems.

(b) LMFBR fuel is generally enriched with plutonium, which presents a different radiological hazard from that of iodine and krypton fission products.

(c) The LMFBR plant is cooled with sodium, therefore water systems cannot be used for pressure reduction or for emergency cooling.

(d) A soluble reactivity poison for control purposes is impracticable in a liquid metal system, since no suitable injection systems exist at the present time.

(e) The LMFBR has a different set of reactivity coefficients from a LWR, and therefore the criteria dealing with coefficients deserve different treatment.

(f) The LMFBR has a different range of accidents and the design basis accidents are very different ones from those in the LWRs. The AEC criteria refer specifically to the loss-of-coolant (as opposed to loss-of-cooling) accident, which is so critical in the LWR systems.

To clear the charge of inapplicability, a set of fast reactor criteria will necessarily be produced within a few years. At that time there are some omissions that could be rectified; the most outstanding being some classification of accidents, their probability ranges and their allowable consequences. This classification could be stated in general terms, whereas at present a classification is only implied in terms of the redundancy required, the emphasis placed on certain accidents, and the definition of an anticipated operational occurrence (an unlikely fault as defined in Section 3.1.3.2).

3.2.3.1 *Examples*

Examples are given to show the need for fast reactor system design safety criteria.

(a) Because of the preoccupation with a high pressure LWR system, the present criteria numbered 14 and 30–32 deal with the *integrity of the coolant boundary*. While still applicable to an LMFBR, the emphasis is wrongly placed.

(b) Criterion 50 dealing with the *containment design basis* states that the containment sustains the loss of coolant accidents. In an LMFBR this is a mild accident, and worse design basis accidents are provided by a core melt-down initiator or a sodium fire (Chapter 5).

(c) Criteria 38–40 deal with *containment heat removal systems* which are very important in the LWR pipe rupture accident when containment pressures might reach 50–60 psig in 10 sec and any reduction in this would alleviate leakage. These systems are not applicable to the LMFBR where the worst accident containment pressures are far less.

(d) Section 3.1 has outlined different classifications of *fuel damage* and some fuel damage during operation may be operationally acceptable. This has to be specifically defined for each plant design. Nevertheless the criteria are far too vague. Criterion 10 states that the design shall "assure that specified acceptable fuel design limits are not exceeded."

The absence of general fast reactor criteria for safety has caused each designer to provide his own set of criteria derived largely by reference to the AEC criteria as guidelines.

3.2.4 PRINCIPAL CRITERIA

The main criteria are those which relate to the release of radioactivity to the environment, 10 CFR 20 and 10 CFR 100, which refer to sections of the Federal Register. They are detailed in Chapter 5.

The 1967 General Criterion 70 states:

Control of Releases of Radioactivity to the Environment. The facility design shall include those means necessary to maintain control over the plant radioactive effluents whether gaseous, liquid or solid. Appropriate holdup capacity shall be provided for retention of gaseous, liquid or solid effluents, particularly where unfavorable environmental conditions can be expected to require operational limitations upon the release of radioactive effluents to the environment. In all cases, the design for radioactivity control shall be justified (a) on the basis of 10 CFR 20 requirements for normal operations and for any transient situation that might reasonably be anticipated to occur, and (b) on the basis of 10 CFR 100 dosage level guidelines for potential reactor accidents of exceedingly low probability of occurrence.

This has been replaced by the 1971 General Criterion 60:

The nuclear power unit design shall include means to maintain suitable control over radioactive materials in gaseous and liquid effluents and in solid wastes produced during normal reactor operation, including anticipated operational occurrences. Sufficient holdup capacity shall be provided for retention of gaseous and liquid effluents containing radioactive materials, particularly where unfavorable site environmental conditions can be expected to impose unusual operational limitations upon their release to the environment.

Regulations 10 CFR 20 and 10 CFR 100 still apply by law although they have been omitted from the wording of the latest criterion. They are the minimum requirements for "suitable control."

Other principal criteria deal with redundancy in protective systems and emergency safety features and these are dealt with specifically in Section 3.3. These criteria too are subject to debate.

3.2.5 DESIGNERS' CRITERIA

A set of very general design criteria is useful in licensing by providing a check list for licensing authorities; however, it does not give the designers any design guidance because there are many different ways of satisfying the criteria. So to further these general rules, a set of *Specific Design Criteria* must be prepared, a number to each of the general criteria.

These specify how the particular system will do the job and they might even enlarge on the protection provided. For example, the specific criteria for number 60 will detail separate gaseous, liquid, and solid waste criteria for low, medium, or high radioactivity levels during normal operation and during accident conditions. These criteria will specify numerical levels of radioactivity in various storage locations and even possibly detail volumes and rates of effluent release. These are then immediately useful to the designers and they provide the safety engineer with a detailed yardstick for measuring the safety of the design against his own safety evaluation.

Of course the safety engineer will also have to satisfy the AEC that the specific set of criteria do indeed satisfy the intent of the general set.

3.3 The Principle of Redundancy

The application of redundancy is an attempt to decrease the probability that an accident will occur.

3.3.1 REDUNDANCY

In reactor systems, a component or system which is vital to the safety of the plant is made redundant; that component or system is doubled in the design, so that in a fault condition one of the components or systems could fail and there would still be one to do the vital job. Thus the probability of failure of the component or system is reduced. There are, however, three main drawbacks to the principle of redundancy.

(a) There is a design inhibition against redundancy; the incorporation of a system or component which may never be used. Some designers may tend to believe one Webster definition of the word redundancy: "an act of needless repetition."

(b) Mere redundancy does not ensure greater safety if the redundant components or systems share vital components or lines, interfere with each other so that a failure of one could cause the other to fail, or are subject to common mode failure (failure from the same fault). Thus true redundancy also implies *independence*.

(c) The cost of doubling up some systems and components is often considerable and may be more than the plant can economically allow.

Thus the application of redundancy requires a certain engineering and safety judgment to obtain optimum safety and operational capability.

3.3.2 APPLICATION

Redundancy is expressed many times in the general design criteria and many times redundancy and independence are tied in with each other (*9*):

Criterion 6: *Sharing of Structures, Systems, and Components*

Structures, systems, and components important to safety shall not be shared between nuclear power units unless it is shown that their ability to perform their safety functions is not significantly impaired by the sharing.

Criterion 17: *Electric Power Systems*

...The onsite electrical power sources, including the batteries, and the onsite electrical distribution system, shall have sufficient independence, redundancy, and testability to perform their safety functions assuming a single failure.

Two physically independent transmission lines, each with the capability of supplying electrical power from the transmission network to the switchyard, and two physically independent circuits from the switchyard to the onsite electrical distribution system shall be provided. Each of these circuits shall be designed to be available in sufficient time following a loss of electrical power from all other alternating current sources, including onsite electrical sources, to assure that specified acceptable fuel design limits and design conditions of the reactor coolant pressure boundary are not exceeded....

Criterion 21: *Protection System Reliability and Testability*

...Redundancy and independence designed into the protection system shall be sufficient to assure that (1) no single failure results in loss of the protection function and (2) removal from service of any component or channel does not result in loss of the required minimum redundancy unless the acceptable reliability of operation of the protection system can be otherwise demonstrated....

Criterion 24: Separation of Protection and Control Systems

The protection system shall be separated from control systems to the extent that failure of any single-control system component or channel, or failure or removal from service of any single protection system component or channel which is common to the control and protection systems leaves intact a system satisfying all reliability, redundancy, and independence requirements of the protection system. Interconnection of the protection and control systems shall be limited so as to assure that safety is not significantly impaired, considering the possibility of systematic, nonrandom, concurrent failures of control system components or channels, or of those common to the control and protection systems.

Criterion 26: Reactivity Control System Redundancy and Capability

Two independent reactivity control systems, preferably of different design principles and preferably including a positive mechanical means for inserting control rods shall be provided.

Then for certain safety features the same phrase occurs in each criterion with only very slight variations in wording. The criteria affected are:

Criterion 34: Residual Heat Removal
Criterion 35: Emergency Core Cooling
Criterion 38: Containment Heat Removal
Criterion 41: Containment Atmosphere Clean-Up
Criterion 44: Cooling Water

and the common phrase is:

Suitable redundancy in components and features, interconnections, and leak detection, isolation, and containment capabilities shall be provided to assure that for onsite and offside electrical power system operation the system safety function can be accomplished assuming a single failure.

In each case it must be remembered that the safety feature or system referred to will not be in operation, unless another fault has already occurred. Thus the single failure to be assumed is an additional single failure.

The addition of this phrase and the definition of single failure is perhaps the biggest change between the 1967 and 1971 sets of AEC criteria.

In these criteria the single failure referred to is defined as follows:

A *single failure* means an occurrence which results in the loss of capability of a component to perform its intended safety functions. Multiple failures resulting from a single occurrence are considered to be a single failure. Mechanical and electrical systems are considered to be designed against an assumed single failure if neither (1) a single failure of any active component (assuming passive components function properly) nor (2) a single failure of any passive component (assuming active components function properly) results in a loss of capability of the system to perform its safety functions. The failure of a passive component need not be considered in the design of mechanical systems if it can be demonstrated that the design is acceptable on some other defined basis, such as an appropriate combination of unusually high quality, high strength or low stress, inspectability, repairability, or short term use (9).

A further example of redundancy is illustrated by the different application of isolation valves:

Criterion 56: *Containment Pressure Boundary Isolation Valves*

Each line which connects directly to the containment atmosphere and penetrates primary reactor containment shall be provided with one automatic isolation valve inside and one automatic isolation valve, other than a simple check valve, outside of containment, unless it can be demonstrated that the design is acceptable on some other defined basis. The valve outside the containment shall be located as close to the containment as practical and upon loss of actuating power the automatic isolation valves shall be designed to take the position that provides greater safety.

With these examples it is clear that redundancy is an important concept to the safety engineer and the design engineer.

3.3.3 EXAMPLES OF REDUNDANCY

Redundancy in valves that are normally open is provided by valves in series, and in valves that are normally closed by valves in parallel. A valve which has no safety function is not required to be redundant even though it provides an extra degree of safety.

Redundancy in power supplies is provided by parallel supplies: onsite power is backed by offsite power, by diesel generators and ultimately by battery power if needed.

Separation in protection systems includes a physical separation of cable lines to avoid jointly disabling cables by electrical-short-induced cable fires and other hazards. Additional components are sometimes required because the shared component might be put out of action by a fault in one of the independent lines. This is expressed in Criteria 6 and 21.

In addition the single-failure criterion will force safety features to have redundant active components as well as protection against the passive failure. However, the passive failure may not be needed if the mechanical system is of "unusually high quality." This implies a degree of overdesign and thus a low-pressure sodium system may be able to qualify whereas a high-pressure steam loop would find it difficult to qualify for exemption from the subsequent passive failure accident.

3.3.4 PARADOXES

Redundancy with systems or components of different principles may force a designer to adopt a second system simply because it has a different principle, even though it is not as reliable or rapid or as safe as a second system of the same principle would be.

For example, two reactivity control systems provided according to Criterion 26 might comprise a group of normal absorber rods together with a poison injection system with speed, distribution, and effectiveness problems. This system would not be as safe as two rod systems, especially if they use different drive mechanisms. However, in the two rod system case, the designer would have to ensure that no common mode failure (such as a core distortion) were possible.

The principle of requiring an engineered safety feature to operate after a fault even in the event of an active component failure at any time, has been extended to requiring an engineered safety feature to operate after a fault even in the event of a passive component failure (see Section 3.3.2).

Thus, for example, an emergency cooling system which goes into operation after a pump failure must be able to withstand subsequently a failure of an active component (valve or even part of its own pump), or a failure of a passive component (circuit or pipe wall or an electromagnetic pump component). This is an extension of the original active component failure criterion of the 1967 set.

So far it is not suggested that the engineered safety features must withstand both an active and a passive failure after the original fault. However one of the paradoxes of criteria is that they grow. It is important to strike a balance between the maintenance of safety and an unreasonable limitation on the design. Overredundancy results in complex systems which are often less amenable to maintenance and are possibly less safe than the simpler less redundant system which is easier to design and test. For safe design it is important not to rely on redundancy as a cure-all.

3.4 Safety Features

A safety feature is defined as a system or a method designed to prevent, mitigate, and/or control the consequences of an accident or accidents.

3.4.1 PLACE IN SAFETY ANALYSIS

Section 1.6 has shown that the addition of a single safeguard can remove sections of a fault tree and therefore safety features arise in safety analysis both as preventative and curative mechanisms.

The fault tree of Fig. 1.31 shows several types of safety features: (a) the design feature used for prevention, which includes flow passages designed to be too large and too numerous to be blocked by an obstruction; (b) the added protective system used for mitigation of the accident such as safety instrumentation designed to halt the fault before the incident proceeds to failure so that the plant is shut down and made safe; and (c) the consequence limiting system used for controlling the accident, with a core catcher, a terminal cooling device or containment, which is never intended for use but which is included because the consequences of an uncontrolled incident are so bad.

The purpose of a safety evaluation is, in part, to discover the need for safety features, to set their functional requirements and then to assess their possible operation in the given design in the event of an accident.

3.4.2 PROTECTIVE SYSTEM

As can be seen from Fig. 1.31 the plant protective system (PPS) which detects a failure and shuts down the reactor is the most important safety system. In conjunction with the emergency core cooling system it provides protection against almost all faults. Section 3.1.5 has already outlined a number of trip signals and the trip values which might be used for a typical plant.

3.4.2.1 *Scram Function*

The prime function of the protective system is to ensure fast and reliable scram in response to a trip signal. To ensure that scram is obtained, the principle of redundancy is used, but to avoid spurious scrams, coincidence techniques are employed.

The logic of protective system action is as follows:

(a) A system acting on one signal from one monitor provides a minimum actuation but it does not provide safety against a failure in the single detection or trip line.

(b) A system acting on one out of two trip lines provides redundancy against a single failure.

(c) A system acting on two out of three trip lines provides redundancy and coincidence and so protects against a spurious signal.

(d) A system acting on two out of four trip lines provides for one channel to fail or to be down for maintenance and still provides total safety.

Table 3.6 shows the scram channel redundancies and coincidences for a number of fast reactors. It can be seen that there is a divergence of opinion as to the correct way to instrument a reactor. Notice that EBR-II provides more trips in total although with less redundancy in some than the Fermi reactor.

TABLE 3.6

REACTOR SAFETY SYSTEM: EXAMPLES OF CHANNEL REDUNDANCY AND COINCIDENCE TECHNIQUES[a]

Trip	EBR-II	Dounreay	RAPSODIE	Fermi
Nuclear:				
Period: source range	2 of 3	2 of 3	1 of 3	1 of 2
intermediate range	2 of 3	1 of 2	2 of 3	2 of 3
power range		2 of 3		
Power level	2 of 3	2 of 3	2 of 3	2 of 3
Negative rate of change of power				2 of 3
Thermal:				
Flow: core inlet	1 of 2	1 of 1		
blanket inlet	1 of 2	1 of 1		
reactor outlet	1 of 1			2 of 3
core outlet		2 of 3	2 of 3	2 of 4
Temperature: upper plenum	2 of 4			
core outlet	1 of 1	2 of 3	2 of 3	2 of 4
bulk sodium	2 of 4			
Power-to-flow ratio			2 of 3	
Upper plenum pressure	1 of 1			
Bulk sodium level	1 of 1			
Other:				
Loss of power to pump	1 of 1			
Gas blanket pressure	1 of 1	2 of 3		
Seismograph	1 of 1			

[a] See Yevick and Amarosi (*10*).

Reactor scram in the fast system is accomplished by one of several methods: adding absorber material (Fermi), removing fuel material (DFR and EBR-II), and removing reflector material (CLEMENTINE).

The absorber is either boron carbide or tantalum. The former generates helium and requires replacement, while tantalum decreases the breeding by softening the spectrum, although it does increase the Doppler coefficient. The rod control drives are sometimes spring assisted either to increase the rate of fall throughout the fall or simply to give it an initial acceleration.

TABLE 3.7

FERMI CONTROL-ROD DESIGN PARAMETERS[a]

Parameter	Safety rods	Operating control rods
General		
Reactor power (MWt)	200	200
Guide tube coolant flow (gal/min)	27	39
Rod coolant flow (gal/min)	11	29
Coolant temperature rise (°F)	90	110
Rod life (yr)	8.9[b]	0.6[b]
Poison material		
^{10}B contained (gm/rod)	535	88
B_4C volume (cm³/rod)	554	158
^{10}B enrichment (at%)	57	32
^{10}B burn-up (%)	7[c]	10[c]
Gas release (liters/rod) (STP)	3.5[b]	6.6[b]
Maximum B_4C temperature (°F)	1000	1100
Poison containment tube		
Design temperature (°F)	1200	1200
Maximum wall temperature (°F)	700	750
Thermal stress in tube (psi)	4000	8000
Internal pressure at end-of-life (psi)	660[d]	430[d]
Pressure stress at end-of-life (psi)	6800[d]	2400[d]

[a] See Yevick and Amarosi (10).

[b] Based on 10% ^{10}B burn-up.

[c] Limited by stress.

[d] Based on ASME Unfired Pressure Vessel Code where allowable fiber stress at 1200°F is 6800 psi.

Table 3.7 shows the characteristics of the Fermi control rods and Fig. 3.4 shows the reactivity change as a control rod is inserted. No reactivity change is experienced for 0.35 sec. This includes a trip delay time and an initial rod insertion time for the end of the control rod to reach about a third of the way into the core. The peak reactivity change is felt by the time the end

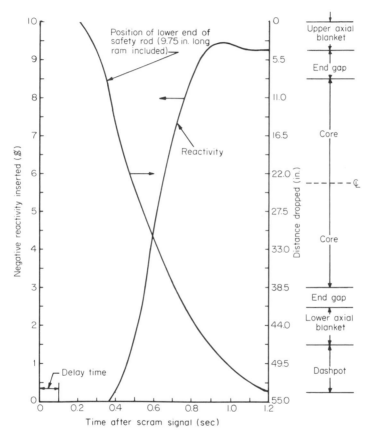

Fig. 3.4. Reactivity inserted as a function of time after scram for the eight-rod scram at Fermi (*10*).

reaches the bottom of the core. The time dependence of the reactivity insertion is the usual S-shaped curve which is taken into account in transient studies.

Table 3.8 shows the comparison of safety rod drive systems in Fermi, EBR-II, and DFR.

TABLE 3.8

COMPARISON OF FERMI, EBR-II, AND DOUNREAY FAST REACTOR CONTROL AND SAFETY ROD DRIVE SYSTEMS[a]

Feature	Fermi	EBR-II	Dounreay
Method of control	Central poison	Peripheral fuel Central fuel backup	Peripheral fuel Peripheral poison backup
Number and type of control rods	10 rods (8 safety, 2 control)	14 rods (12 peripheral control, 2 safety)	12 rods (2 safety, 4 shutoff, 6 control) 3 boron poison backup
Total reactivity $\delta k/k$	More than 0.063	0.063–0.068	More than 0.09
Shield plug design	Single rotating	Double rotating	Double rotating
Coolant flow direction	Up	Up	Down
Drive location	On plug, in line with rods	On plug, in line with rods	Outside plug, offset actuator for rods
Method of connection of drive to rod	Direct, relatively flexible connection	Direct, relatively tight connection	Located on carrier mating cone and pin
Stroke	Safety 54 in., control 20 in.	14 in.	25 in.

Scram method	Safety rods dropped, drive follows fast to assure scram. Spring assisted	All control scram, pneumatic assisted. Safety rods only scram during start-up and refueling	All rods scram. Control dropped with their drives, boron dropped with makeup piece only
Actuation	Electromagnetic latch	Electromagnetic latch	Electromagnetic latch
Scram time total	About 0.9 sec	About 0.32 sec	About 0.5 sec
Type of drive shaft	Electric motor—driving ball nut and screw (external)	Electric motor—driving rack and pinion (external)	Electric motor—gear to ball nut and screw (internal)
Position indicator	Digital readout gear driven	Selsyn system from pinion shaft	Special system from servo-armature and search coil
Speeds (in./min)	Safety: 1.6 out, 120 in. Shim: 0.4 in./out Regulator: variable 1–10	Fixed at 5 in./out	Fixed at 0.18 out, 0.18 or 9 in Boron rods: 0.36
Sealing	Metal O-rings and reciprocating metal bellows	Aluminum gasket and reciprocating metal bellows	O-rings or other metal gaskets, no bellows. All seals static

[a] See Yevick and Amarosi (*10*).

3.4.2.2 *Power Reduction Function*

Some reactor conditions require a reduction in power. This can be accomplished safely through the scram function but it may be not necessary or desirable to scram in every case. A series of partial power reductions is sometimes more appropriate. In sequence: (a) warning, (b) setback in power, (c) controlled shut-down, and (d) scram. Such a scheme would avoid thermal shocks to the reactor components in many instances.

Table 3.9 shows the standard earthquake scale (*11, 12*). Because of the possibility of structural movement bringing about core reactivity changes, the reactor may require power reductions when a serious earthquake occurs. Generally a scram is initiated at a modified Mercalli intensity of V. However, it is more consistent with modern operating practice and modern knowledge of seismology to provide a power setback system on the following scheme (*11*).

At a modified Mercalli intensity of V a warning is provided, at an intensity of VII a controlled shut-down would be instituted and then at an intensity of VIII a scram would be initiated.

This philosophy can be applied to many secondary warning indications in the plant without sacrificing any of the overall safety of the system. Usually, however, the warning and setback functions are part of the control system rather than the protective system, and it is necessary to coordinate the two systems most carefully to ensure overall safety under this scheme.

3.4.3 OTHER SAFETY FEATURES

It is appropriate to discuss safety features in relation to the accidents which they were designed to mitigate.

3.4.3.1 *Prevention of Loss of Coolant*

Following a break in the primary circuit the following engineered safety features might be invoked by a particular design. (Naturally not all of them would be required and exactly which would be used would be a function of the plant size, layout and concept details, and other safety features being used.)

Plant protective system to shutdown the reactor.

Wrapped and restrained pipework to guard or restrict pipe breakage.

Safety guard tanks around components (or minimum volume containers to restrict coolant loss).

Double pipework to avoid coolant loss.

TABLE 3.9

COMPARISON OF EARTHQUAKE SCALES[a]

ROSSI–FOREL INTENSITY SCALE (1883)	MODIFIED–MERCALLI INTENSITY SCALE (1931), WOOD AND NEUMANN	GROUND ACCELERATION a	MAGNITUDE M	ENERGY E	EPICENTRAL ACCELERATION a_o
COL 1	COL 2	COL 3	COL 1	COL 2	COL 3
	I Detected only by sensitive instruments	$\frac{cm}{sec}\ \frac{\alpha}{g}$		Ergs	$\frac{cm}{sec}\ \frac{\alpha_o}{g}$
I The shock felt only by experienced observer under very favorable conditions	II Felt by a few persons at rest, especially on upper floors; delicate suspended objects may swing	2		10^{14}	2
II Felt by a few people at rest; recorded by several seismographs	III Felt noticeably indoors, but not always recognized as a quake; standing autos rock slightly, vibration like passing truck	3, 4, 5 .005g, 6, 7	M–3	10^{15}, 10^{16}	3, 4, 5 .005g, 6, 7
III Felt by several people at rest; strong enough for the duration or direction to be appreciable	IV Felt indoors by many, outdoors by a few; at night some awaken; dishes, windows, doors disturbed; motor cars rock noticeably	8, 9, 10 .01g	M–4	10^{17}	8, 9, 10 .01g
IV Felt by several people in motion; disturbance of movable objects, cracking of floors	V Felt by most people; some breakage of dishes, windows, and plaster, disturbance of tall objects	20, 30		10^{18}	20
V Felt generally by everyone, disturbance of furniture, ringing of some bells					
VI General awakening of those asleep, ringing of bells, swinging chandeliers, startled people run outdoors	VI Felt by all; many frightened and run outdoors; falling plaster and chimneys; damage small	40, 50 .05g, 60, 70, 80, 90, 100 .1g	M–5	10^{19}	30, 40, 50 .05g, 60, 70, 80, 90, 100 .1g
VII Overthrow of movable objects, fall of plaster, ringing of bells, panic with great damage to buildings	VII Everybody runs outdoors; damage to buildings varies, depending on quality of construction; noticed by drivers of autos			10^{20}	
VIII Fall of chimneys; cracks in walls of buildings	VIII Panel walls thrown out of frames; fall of walls, monuments, chimneys; sand and mud ejected; drivers of autos disturbed	200	M–6		200
IX Partial or total destruction of some buildings	IX Buildings shifted off foundations, cracked, thrown out of plumb; ground cracked; underground pipes broken	300, 400, 500 0.5g, 600, 700, 800, 900 1g, 1000	M–7	10^{21}, 10^{22}	300, 400, 500 0.5g, 600, 700, 800, 900 1g, 1000
X Great disasters, ruins; disturbance of strato, fissures, rockfalls, landslides; etc.	X Most masonry and frame structures destroyed; ground cracked; rails bent; landslides				
	XI New structures remain standing; bridges destroyed; fissures in ground; pipes broken; landslides; rails bent	2000, 3000	M–8	10^{23}	2000, 3000
	XII Damage total; waves seen on ground surface; lines of sight and level distorted; objects thrown up into air	4000, 5000 5g, 6000		10^{24}	4000 5g

RICHTER SCALE

[a] See Lomenick (*11*).

Elevated design and layout designed to minimize outflow.
Check valves to reduce loss of coolant.
Hydraulic diodes to reduce loss of coolant.
Reserve coolant volumes to provide for coolant loss.
Siphon breakers to reduce outflow.
Isolation valves to isolate the failed loop or pipe.
Multiple path design to minimize the effect of a pipe rupture.

3.4.3.2 *Prevention of Local Core Damage*

Following local damage it is important to contain the existing damage as locally as possible and to prevent the spread of damage. One must consider a plant protective system, multiple path inlets to assemblies to avoid blockage, strong assembly canning to contain any local failure, and emergency core cooling system (and pony motors on main pumps) to avoid failure after loss of cooling.

3.4.3.3 *Prevention of Criticality Following Fuel Melt-down* (*Consequence Limiters*)

The core fuel after fuel melt-down would need to be prevented from accumulating into a critical mass (see also Section 5.6.1). Preventatives could include large heat sinks at subassembly inlets to freeze molten fuel from single subassembly, dispersion cones in and out of vessel, melt catchers (assembly or core size), terminal cooling systems, or ground disposal to permanently avoid criticality.

3.4.3.4 *Containment Following a Core Dispersion Criticality* (*Consequence Limiters*)

After a core dispersion accident, it would still be necessary to contain the fission gas and any plutonium in the system as far as possible (see also Sections 5.5 and 5.6.2). The following methods might be used: radial blast shielding, shock attenuation volumes, sodium hammer suppressor plate, head hold-down system, missile barriers, splash plates to agglomerate sodium sprays, aerosol settling devices, filtration, pressure suppression systems, and containment barriers.

3.4.3.5 *Prevention of Sodium Reactions*

The easiest way to prevent chemical reactions is to prevent the constituents from coming into contact. One would need inerted vaults and containment, absence of water in system, and drainage points for sodium spills.

It should be reiterated that the main safety feature is that the fault should always be detectable and then that remedial action can be taken by the protective system: to scram and to provide emergency core cooling. The criteria (9) which apply to safety features have been discussed in Section 3.3.

REFERENCES

1a. C. P. Gratton, E. G. Bevan, J. Graham, R. Hobday, and A. T. Hooper, A gas-cooled fast reactor using coated particle fuel. *J. Brit. Nucl. Energy Soc.* **7** (3), 233 (1968).

1b. H. Lurie, Steady state sodium boiling and hydrodynamics. NAA-SR-11586, Atomics Int., Canoga Park, California, January 15, 1966.

2a. O. D. Kazachkovskii, S. N. Votinov, I. G. Lebedev, *et al.*, Study of the operation of an assembly of fuel elements containing plutonium dioxide fuel in the BR-5 Reactor. *At. Energ.* **24**, 136–43 (1968); translated as ANL-Trans-609 (1968).

2b. K. Bagley, ed., Void formation in cladding and structural materials under irradiation. Brit. Rept. TRG-Memo-4950 (Revised), U. K. Atomic Energy Authority, August 1969.

3a. B. R. T. Frost, Theories of swelling and gas retention in ceramic fuels. *Nucl. Appl.* **9** (2), 128 (1970).

3b. M. J. F. Notley, A computor program to predict the performance of UO_2 fuel elements irradiated at high power outputs to a burnup of 10,000 MWd/MTU. *Nucl. Appl.* **9** (2), 195 (1970).

4. J. E. Hanson, J. H. Feld, and S. A. Rabin, Experimental studies of transient effects in fast reactor fuels, Ser. II. Mixed oxide $(PuO_2–UO_2)$ irradiations. GEAP-4804, General Electric Corporation, San Jose, California, May 1965.

5. M. D. Carelli, Fuel rod design limit and transient survival criteria. WARD-4135-6, Westinghouse Advanced Reactors Division, Waltz Mill, Pennsylvania, May 1970.

6. G. S. Budney, Liquid-metal-heated steam generator operating experience. NAA-SR-12534. Liquid Metal Eng. Center, Atomics Int. Canoga Park, California, November 1967.

7. "Failure Data Handbook for Nuclear Power Facilities," Vol. I. Failure data and applications technology, LMEC-Memo-69-7, Vol. I. Liquid Metal Eng. Center, Atomics Int. Canoga Park, California, 1969.

8. General design criteria for nuclear power plant construction permits, Appendix A, 10 CFR 50. Code of Federal Regulations, Fed. Regist., July 11, 1967.

9. General design criteria for nuclear power plants, 10 CFR 50. Code of Federal Regulations, Fed. Regist., February 10, 1971.

10. J. G. Yevick and A. Amarosi, eds., "Fast Reactor Technology: Plant Design." M. I. T. Press, Cambridge, Massachusetts, 1966.

11. T. F. Lomenick, and NSIC Staff, Earthquakes and Nuclear Power Plant Design. ORNL-NSIC-28. Oak Ridge Nat. Lab., Oak Ridge, Tennessee, July 1970.

12. C. F. Richter, "Elementary Seismology." Freeman, San Francisco, California, 1968.

CHAPTER 4

SPECIAL FAST REACTOR CHARACTERISTICS

4.1 Heat Ratings and Coolants

The fast reactor core is typically much smaller than a thermal reactor system with the same heat output, since there is no requirement for a moderating material. Much higher heat ratings result from the size of the core, and the problem of removing heat from the system becomes a major design problem.

In order to provide more heat transfer surface for the heat removal process, the fuel is subdivided more than it is in a thermal system. Pins will have characteristically about half the diameter of the thermal reactor. Figure 4.1 illustrates the difference in size of fuel pin lattice and core between a 1000 MWe PWR and a 1000 MWe LMFBR (*1*).

The fast reactor is very small with 10 times the volumetric PWR heat rating. Thus an extremely efficient coolant is required for heat removal from the very close lattice that is inherent in the fast reactor.

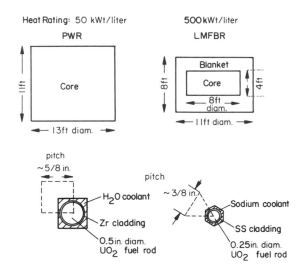

Fig. 4.1. Comparison of characteristic dimensions for 1000 MWe PWR and LMFBR Cores (*1*).

4.1.1 CHOICE OF FAST REACTOR COOLANTS

The choice of a coolant is defined by a number of practical considerations. Primarily, excellent heat transport properties are required, but, in addition, the coolant should have little chemical activity and be compatible with the containment material so that corrosion is minimized. It should be stable and not be subject to decomposition even under irradiated conditions; its melting temperature should be low to avoid having to preheat the system to obtain a liquid coolant, but its boiling temperature should be high to avoid having to impose high pressures to suppress boiling. In addition its nuclear properties should be such that it is a poor moderator, and its induced radioactivity build-up should be low to avoid having excessive shielding around the primary system. Added to all these requirements is the final one, that the coolant should be inexpensive; at least its cost should be a small fraction of the total cost of the plant.

Table 4.1 lists coolant properties for 15 potential liquid-metal coolants for fast reactors together with three evaluation parameters with which the effectiveness and suitability of a coolant candidate may be judged (*2*).

(a) *The pumping power.* This factor is a measure of the efficiency of the coolant. For systems with the same geometry, the same heat output, and the same temperature rise in the coolant channel, the pumping power varies as

$$\mu^{0.2}/\varrho^2 c_p^{2.8}$$

In this expression, μ is the viscosity in lb/ft-hr, ϱ is the density in lb/ft³, and c_p is the specific heat in Btu/lb-°F. The pumping power is the product of the coolant volume flow and the channel pressure drop which is proportional to the friction factor. The friction factor is proportional to $R_e^{-0.2}$.

A low value of the pumping power is required in order to avoid an economic penalty in heat transport. Table 4.1 shows that lithium has the lowest pumping power factor followed by zinc and gallium. Sodium ranks fourth in this assessment.

(b) *Induced activity*. The buildup in induced reactivity increases the amount of shielding required for the primary system and it therefore should be kept as low as possible. The specific activity in Ci/gm is given by

$$S = \frac{1}{K} \sum_{ij} \frac{f_i N_0 \varphi_j}{A_w \tau_0} \sigma_{ij} \tau_r [1 - \exp(-0.6930/\theta_i)] \tag{4.1}$$

where f_i is the atomic number of the ith coolant isotope; N_0 is Avogadro's number $(0.60248 \cdot 10^{24})$; φ_j is the neutron flux in the jth energy interval $(n/cm^2\text{-sec})$; σ_{ij} is the microscopic cross section for the ith isotope and the jth energy interval $(cm^2/nucleus)$; A_w is the coolant atomic weight (gm/gm-mole); τ_r is the coolant residence time within the neutron flux for one coolant cycle (sec); τ_0 is the cycle time for the coolant (sec); θ is the irradiation time (sec); θ_i is the half-life for the ith isotope (sec); and K is $3.7 \cdot 10^{10}$ disintegrations/Ci-sec.

The exponential term accounts for the loss of activity due to the fast decay. Note too that the two cycle times τ_r and τ_0 depend on the heat transfer properties.

Table 4.1 once again shows that lithium is the most suitable coolant based on this assessment, followed by lead and an alloy of lead and bismuth. Sodium ranks seventh in the list although the addition of potassium brings it into fourth position.

(c) *A temperature range ratio*. In order to obtain some quantitative assessment of the working range of the coolant the following ratio of temperatures provides a useful measure

$$(T_b - T_m)/T_w \tag{4.2}$$

T_b, T_m, and T_w are, respectively, the boiling, melting, and outlet absolute temperatures of the coolant (°R).

TABLE 4.1

Property	Bismuth	Gallium	Lead	Lithium (nat.)	Mercury
General physical properties					
Atomic number	83	31	82	3	80
Atomic weight	209	69.72	207.21	6.94	200.61
Density, lb/cu/ft	608	359	650	29.9	826[c]
Viscosity, lb/ft/hr	2.664	1.836	4.1	0.82	2.76[c]
Surface tension, lb/ft	0.0247	NA	0.0293	0.0238	0.0307[c]
Resistivity, $\mu\Omega$/cm	142.78	NA	104	36	108.3[c]
Vapor pressure, Torr	Neg.	Neg.	Neg.	Neg.	9300
Thermal properties					
Melting point, °F	520	85.86	621	357	−37.97
Boiling point, °F	2691	3601	3159	2428	675
Specific heat, Btu/lb/°F	0.0369	0.082	0.0346	0.996	0.0326[c]
Thermal conductivity, Btu/hr/ft/°F	8.95	18.0	8.84/0.91	27.6	5.8[c]
Heat of fusion, Btu/lb	21.6	34.49	10.60	185.9	5.0
Heat of vaporization, Btu/lb	367.74	1825	368.3	8349	125
Prandtl number ($c_p\mu/k$)	0.011	0.0083[f]	0.016	0.0295	0.0154[c]
Vol. change on fusion, % solid vol.	−3.32	−3.1	3.6	1.5	3.6
Nuclear properties					
Fast activation cross section, mb	0.6	28.0	3.1	0.03	89.0
Nonelastic scattering cross section at 2 MeV, b	0.53	NA	0.64	<1	2
n, γ daughter half-life	None	14.1 hr	None	None	48 days
n, γ daughter activity, MeV	None	0.84, 0.60, 3.35	None	None	0.28
ξ, av log energy decrement	0.0096	0.0287	0.0097	0.2643	0.0100
General properties					
Cost/lb (as of 1960)	$ 2.25	$ 675	$ 0.12	$ 11[k]	$ 2.82
Container material	Cr-Mo steel	Graphite	Carbon steel	Cb-Zr	Ferrous metals
Reactions with U	Soluble	Soluble	Sl. sol.	Slight	Soluble
Reaction with Pu	Soluble	Soluble	Sl. sol.	Slight	Soluble
Toxicity	Slight	Slight	High	Moderate	High
Fire or explosion hazard	Slight	None	Moderate as dust	High	None
Evaluation parameters					
$\mu^{0.2}/(\varrho^2 c_p^{2.8})$, pumping power criterion	0.034	0.009	0.039	0.00109	0.0261
S (specific activity), Ci/gm for 1,000-MW(t) reactor	0.17	2.3	m	n	0.28
$(T_b-T_m)/T_w$, temp. range ratio	1.49	2.41	1.74	1.40	0.94

[a] All properties are for a temperature of 1000°F unless otherwise specified; NA = not available.
[b] Factors for conversion from engineering units to CGS units:

Temperature	°C = (5/9)(°F − 32)
Latent heat	1 Btu/lb (mass) = 0.5556 cal/gm
Surface tension	1 lb (force)/ft = 1.459 × 10^4 dyn/cm
Density (ϱ)	1 lb (mass)/cu ft = 0.01602 gm/cm^3
Specific heat (c_p)	1 Btu/lb (mass)/°F = 1 cal/gm/°C
Viscosity (μ)	1 lb (mass)/ft/hr = 4.134 × 10^{-3} gm/sec, or poise
Thermal conductivity (k)	1 Btu/hr/ft/°F = 4.134 × 10^3 cal/sec/cm/°C
Prandtl number	$N_{Pr} = c_p\mu/k$, dimensionless
Thermal diffusivity $k/(c_p\varrho)$	1 sq ft/hr = 0.2581 cm^2/sec.

PROPERTIES OF LIQUID METALS[a]

Potassium	Rubidium	Sodium	Tin	Zinc	Na(56%)-K(44%)	Na(22%)-K(78%)	Pb(44.5%)-Bi(55.5%)	Sulfur	Phosphorus
19	37	11	50	30	—	—	—	16	15
39.100	85.48	22.997	118.70	65.38	30.082	35.557	208.2	32.066	31
44.6	84.4	51.4	421	428	47.4	46.3	625.5	102.5[d]	84.9[e]
0.41	0.415	0.55	2.736	6.192	0.47	0.43	2.88	680[d]	<1
0.0057[c]	0.0022[f]	0.0101	0.0347	0.0535	0.0071[c]	0.0078[c]	NA	0.00294	NA
48	60[f]	29	55.54	35.22	61	71	129.85	173[d]	NA
62	139	9	Neg.	Neg.	26[f]	45[f]	Neg.	310.2[d]	227.5[e]
146	102	208.1	449	787	66.2	12	257	246	111.5
1402	1270	1618	4118	1663	1518	1443	3038	832	536
0.182	0.0877	0.301	0.0639	0.1165	0.2485	0.209	0.035	0.248[d]	NA
21.2	13.2	37	19.0	33.2	16.4	15.05	8.05	0.0952	NA
25.5	11.0	48.7	26.1	43.92	NA	NA	NA	25 8[g]	9.06[g]
853	363	1718	1031	755.1	NA	NA	NA	121.2	254[g]
0.0035	0.00276[f]	0.0044	0.0095	0.022	0.0071	0.0060	0.0125	1771	<1
2.41	2.5	2.5	2.6	6.9	2.5	2.5	0.0	NA	NA
0.59	46.0	0.87	14.0	13.0	0.80	0.64	3.5	1.4[h,i]	3[h,i]
4(3.7 MeV)	NA	<1	1	1	NA	NA	0.58[j]	NA	NA
12.4 hr	19 days	15 hr	112 days	None	See K, Na	See K, Na	None	None	None
0.32, 1.51	1.1	2.775, 1.368	0.39	None	NA	NA	None	None	None
0.0507	0.0234	0.0852	0.0169	0.0305	NA	NA	NA	0.0616	0.0637
$ 3.66	$ 390	$ 0.17	$ 1.00	$ 0.13	$ 0.60	$ 0.80	See Pb, Bi	$ 0.08	$ 0.09
304 SS	304 SS	304 SS	Quartz	Graphite	304 SS	304 SS	Cr-Mo steel	Graphite	NA
Slight[l]	Slight[l]	Slight[l]	Sl. sol.	Sl. sol.	Slight	Slight	Sl. sol.	Reacts	Reacts
Slight[l]	Slight[l]	Slight[l]	Alloys	NA	Slight	Slight	Sl. sol.	Reacts	Reacts
High	Moderate	High	None	None	High	High	High	Slight	Moderate
High	High	High	Slight as dust	Moderate	High	High	Moderate as dust	Slight	High
0.050	0.107	0.0097	0.015	0.003	0.0189	0.0316	0.024	0.0174	NA
0.11	2.6	0.20	0.95	1.17	0.16	0.11	0.09	0.80	[o]
0.86	0.80	0.97	2.51	0.60	0.99	0.98	1.91	0.90	0.47

[c] At 300°F.
[d] At 732°F.
[e] At 436°F.
[f] Computed from data.
[g] At 15°C.
[h] Effective $\sigma_{n,p}$ over energy spectrum.
[i] (n, p) reaction.

[j] Computed from lead and bismuth.
[k] 99.99% [7]Li, $ 54.50 per lb.
[l] Due to impurities.
[m] No gamma activity; beta activity is 0.090 Ci/gm.
[n] No gamma activity; beta activity from [8]Li is about 0.03 Ci/gm.
[o] No gamma activity; beta activity is about 1.5 Ci/gm.

A high value of this temperature ratio is desirable for a practical coolant, and Table 4.1 shows that tin, followed by gallium and an alloy of lead and bismuth head the merit table. This time lithium comes sixth while sodium comes eighth in the list. Other temperature factors could be used to assess working ranges but they give similar results.

Using these three evaluation parameters to choose a liquid-metal coolant would appear to lead to a choice of lithium followed by gallium.

However, lithium has a strong absorption cross section for fast neutrons. The main culprit is the ^{6}Li content and ^{7}Li comprises 92.5% by weight of natural lithium, so the absorption can be reduced by enriching the lithium in its ^{7}Li content. Lithium also has the disadvantage of having a low atomic number and it is therefore a strong moderator, three times as effective as sodium, which itself is a strong moderator. Strong moderation is undesirable because it degrades the spectrum and lowers the breeding ratio, and since breeding and neutron economy are most important to the success of the fast reactor, lithium cannot be used as a coolant. Gallium is far too expensive, even though the price could be expected to decrease if it were used in bulk.

Sodium is chosen as a coolant therefore, because it is a better-than-middle runner in all the assessment contests.

4.1.1.1 Chemical Activity of Sodium

The following comprise the principal sodium reactions of interest for exposed sodium (3).

$$4Na + 3CO_2 \rightarrow 2Na_2CO_3 + C$$
$$2Na + 3CO \rightarrow Na_2CO_3 + 2C$$

$$Na + H_2O \rightarrow NaOH + \tfrac{1}{2}H_2$$
$$Na + NaOH \rightarrow Na_2O + \tfrac{1}{2}H_2$$

$$Na + \tfrac{1}{2}H_2 \rightarrow NaH$$

$$2Na + \tfrac{1}{2}O_2 \rightarrow Na_2O$$
$$2Na + O_2 \rightarrow Na_2O_2$$

The final two reactions are discussed in Section 4.5, which is concerned with sodium fires. It suffices to say here that air has to be excluded from areas in which sodium is handled and where temperatures may be high. Both these reactions release considerable heat and dense oxide smoke.

Components which have been immersed in sodium are cleaned before inspection or repair. The cleaning is performed on the basis of either dissolving the sodium or reacting with it before rinsing away the cleanser and the cleaning products. Cleaners are alcohol, ammonia, nitric acid or other acids such as hydrofluoric, acetone, and steam.

4.1.2 CHOICE OF COOLANTS OTHER THAN SODIUM[†]

Coolants other than liquid metals are possible and previous sections of this book have dealt with reactor systems using supercritical steam and helium as coolants.

Although the U.S., British, French, and Soviet national fast breeder programs are based on sodium, gaseous coolants have not been ruled out. Indeed the gas-cooled system shows considerable breeding potential. Figure 4.2 shows comparative breeding ratios for different fuels and coolants.

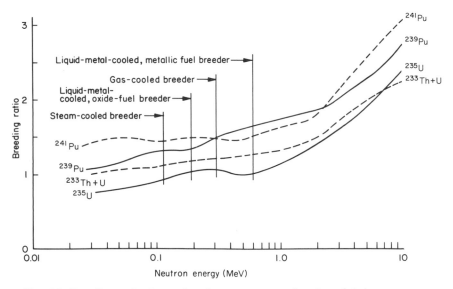

Fig. 4.2. Breeding ratios for various fast reactors as a function of their mean neutron energy (*4c*).

Steam gives rise to a poorer breeding ratio, because it degrades the spectrum while low density helium, although a moderator, actually improves the spectrum in comparison to sodium.

In order to obtain adequate heat removal, a nonmetallic coolant must be used at high pressure: over 3000 psi for supercritical steam and of the order

[†] See Dalle-Donne (*4a*) and Hummel and Okrent (*4b*).

of 1000 psi for the helium gas-cooled system. This high pressure then gives rise to problems associated with depressurization of the system.

Gas-cooled fast reactors are susceptible to water-flooding accidents, either from an external source or from the high-pressure steam generator. The resultant reactivity increase can be prohibited by the addition of resonance absorbers to the system to absorb the thermalized neutrons, but in practice this would probably be unneccessary because flooding-induced reactivity changes would be slow transients and could be engineered out of the system.

Despite the economic potential of gas-cooled fast reactors, the remainder of this volume will refer only to sodium-cooled systems, unless otherwise noted, because sodium is the chosen coolant for the first generation of fast-breeder power reactor plants around the world.

4.2 Voiding Effects

Liquid-metal coolants may contain voids due to externally introduced gases or due to vaporization of the coolant itself. Previous sections have already dealt with some aspects of core voiding and its analysis. Section 1.3 outlined modeling of sodium boiling due to an imbalance of heat production and heat removal in a core channel. Section 1.4 discussed the reactivity feedbacks obtainable from voiding and outlined design choices which were available to reduce these void feedback coefficients. Finally Section 2.3 dealt with integrated reactivity effects and reactivity addition rates from possible bubbles passing through a reactor core. This section discusses the sources of possible external gases in a practical system and investigates what effects bubbles in the core might have, in addition to reactivity feedbacks.

4.2.1 SOURCE OF BUBBLES

In a practical system, there are a number of sources which should be considered both during operation and during accident conditions. For illustrative purposes we consider a loop-type plant.

(a) *Preoperational filling.* It is possible that a considerable number of free spaces may have entrapped gas when the primary circuit was initially filled and these entrapped gases could be swept into the coolant stream if they were allowed to remain. However, before operation, cold and hot hydraulic tests of the system at low power will ensure that such gases are removed from the circuit.

(b) *Gas entrainment.* Cover gases may be entrained from a free surface if a vortex is formed due to the particular flow path of the design. In a loop-type plant the outlet nozzles should be sufficiently immersed to avoid the formation of vortices.

Gas may also be entrained by differences of pressures inside and outside instrument guide tubes that dip below the sodium free surfaces. If the cover gas systems are not regulated well, then it is possible that streams of gas may be introduced within the sodium (*4d*). However in all cases gas entrainment can be prohibited by good engineering design. The problem arises in making sure that all possible cases have been considered. Accident experience has shown that this is not easy.

(c) *Gas absorption at a free surface.* Gas may be collected by a fluid in motion below a cover gas, and this may later be concentrated within the primary circuit at some high point of the system. However absorption rates are very small, of the order of hundredths of a pound of gas per hour; this may be reduced by ensuring that the sodium in contact with cover gas volumes is nonturbulent.

(d) *Fission gas release.* The core does contain gas in each fuel pin fission gas plenum and a sudden release of a lot of this high-pressure gas could give rise to a considerable volume of gas in the circuit. For this reason the fission gas plenum is, in most designs, at the outlet of the core rather than at the inlet, so that any fission gas would be immediately swept out of the coolant channels. However, fuel pins are not likely to rupture in large quantities unless something else is seriously at fault, and gas released from small numbers of pins is rapidly removed from the circuit at the vessel free surface and hot and cold purification traps.

(e) *Production from oil releases.* If a pump lubricant, say the Fluorolube M-10 used in the Fermi pumps, could possibly penetrate to the primary circuit, then in contact with hot sodium, the lubricant would decompose into gaseous products. To avoid this possibility the system is engineered with multiple seals, a tortuous path for any possible leakage, and an oil sump well removed from the sodium coolant. Indeed, if this source were considered possible, then an oil which is compatible with sodium could be used.

(f) *External purification and make-up circuits.* External purification lines have cover gas systems that may give rise to gas sources, and although in any case it is simple to design the system to avoid a source for primary circuit gas, it is nevertheless difficult to ensure that all possibilities have been covered.

(g) *Entry at pipe rupture.* In certain loop designs in some low flow circumstances, it is possible that part of the circuit may be at less than atmospheric pressure. A leak in this region would result in an inleakage of external gas to the primary. Alternatively, a large break may result in an input of gas, even though that break took place at a part of the circuit which was originally at high pressure. Thus, circuits should be designed to avoid inleakage of gas in the remote event of a rupture of the primary circuit.

If, despite engineering precautions to prohibit gas from entering the primary system, gas were to enter, it is unlikely that after passage through the pipework, the heat exchanger, and the pumps that a coherent bubble would result at the core inlet. Tests have shown that any bubble is dispersed by inlet plenum flows rather than concentrated at a single point. Therefore it is most unlikely that externally introduced gases could give serious reactivity effects. However, it *is* likely that very small bubbles will penetrate to the core and their effects on the heat transfer should be considered.

4.2.2 EFFECT OF BUBBLES ON HEAT TRANSFER

Bubbles within coolant channels may affect the heat transfer in two ways. If the bubbles are small and dispersed and spread evenly throughout the coolant, they have a homogeneous effect of macroscopically decreasing the density of the coolant. If the bubbles are larger, they may be considered individually as insulating a portion of the fuel pin either in a stationary manner or in a transient manner as the bubble moves up the core.

4.2.2.1 *Homogeneous Effects*[†]

If the bubbles are small and dispersed, the density is effectively reduced. The Nusselt number is given by an expression of the form

$$N_u = 4.8 + 0.025\,P_e^{0.8} \qquad (4.3)$$

Thus as $N_u = hD/k$ and $P_e = \varrho D v(c_p/k)$ this equation directly leads to an evaluation of the heat transfer coefficient h, which is much simplified if one assumes that the density of gas is negligible compared to the density of sodium.

With gas in the sodium coolant, the effective density is given by Eq. (4.4) in which α is the void fraction. The average velocity of the coolant and the

[†] See Hori and Hosler (*52*).

heat capacity c_p' are given by Eqs. (4.5) and (4.5a).

$$\varrho' = \varrho(1 - \alpha) \tag{4.4}$$

$$v' = v(1 - \alpha)^{-1} \tag{4.5}$$

$$c_p' = c_p(1 - \alpha) \tag{4.5a}$$

For a two-phase mixture the thermal conductivity for small void fractions is given (5a) by Eq. (4.6). This equation is valid up to a void fraction of 0.5

$$k' = k(1 - \alpha)/(1 + \tfrac{1}{2}\alpha) \tag{4.6}$$

Thus substituting these values for k', v', c_p', and ϱ' for the single phase values in Eq. (4.3) the homogeneous coolant heat transfer equation becomes

$$h' = h[(1 - \alpha)/(1 + \tfrac{1}{2}\alpha)][4.8 + 0.025 \, P_{\mathrm{e}}^{0.8}(1 + \tfrac{1}{2}\alpha)^{0.8}] \tag{4.7}$$

In the limit of the validity of this equation, for a void fraction of 50%, the heat transfer is reduced to about a third of its original value.

The fuel temperature is related to the coolant temperature in steady-state conditions through the heat-transfer coefficient as

$$\text{fuel pin surface temp.} = \text{coolant temp.} + \text{power}/h' \tag{4.8}$$

Thus the temperature difference between the surface of the fuel pin and the coolant will increase by something less than three times. Thus, even with a void fraction of 50% the surface temperature on the hot pin cladding might increase from 1200 to 1500°F and failure is unlikely. In practice, this volume of bubbles would not be possible.

4.2.2.2 *Local Effects*

The other possibility is that the bubble in the channel would be relatively large and would either lodge somewhere or would pass through the core transiently insulating the fuel pin as it passes through.

Studies have shown that the passage of single bubbles at normal sodium velocities, assuming complete insulation of the pin surface under the bubble, results in perturbations of tens of degrees rather than hundreds. Thus no sudden failure is likely (5b). If the bubble were to lodge in a single position for a comparatively long time, then the surface temperature could increase, unless the axial conduction or circumferential conduction paths were adequate in removing the heat from the surface under the bubble.

Axial conduction is not efficient and thus most heat is removed circumferentially through the cladding from beneath the bubble. Whether this latter path is efficient or not depends on the circumferential coverage of the bubble. For small bubbles the heat is removed through the cladding faster than it is input from the fuel within the cladding, but for larger bubbles the heat input is the faster process. Thus there is a critical coverage which may be obtained from a comparison of the time constants for heat input into the cladding and heat transport circumferentially through the cladding.

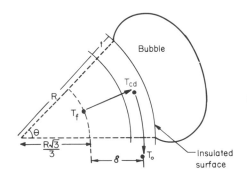

Fig. 4.3. Model for bubble blanketing of a fuel pin surface.

For a bubble which covers an angle θ of the fuel pin surface and assuming no heat removal (see Fig. 4.3), the time constant for the input of heat into the cladding below the bubble is given by

$$\tau_{in} = (\varrho_f c_f \varrho_{cd} c_{cd} R^2 \, \delta t)/(\varrho_f c_f R + 2\varrho_{cd} c_{cd} t) \, K_f (R/\sqrt{3} + \tfrac{1}{2}\delta) \qquad (4.9)$$

Assuming no heat input, the time constant for heat removal along the cladding in a circumferential direction is given by

$$\tau_{out} = (\varrho_{cd} c_{cd} R^2 \theta^2)/4K_{cd} \qquad (4.10)$$

The critical angle of coverage θ_{cr} is defined as that angle when these time constants are equal. Below this angle of coverage the cladding surface temperature rises only moderately above normal values, whereas above this value cladding temperatures rise considerably more. The critical angle θ_{cr} is given by

$$\theta_{cr} = (4\varrho_f c_f K_{cd} \, \delta t)^{1/2}/[(\varrho_f c_f R + 2\varrho_{cd} c_{cd} t) \, K_f (R/\sqrt{3} + \tfrac{1}{2}\delta)]^{1/2} \qquad (4.11)$$

An approximate expression may be derived as

$$\theta_{cr} = 0.268(8K_{cd} t/K_f R)^{1/2} \qquad (4.12)$$

For a fuel pin of 0.25-in. diameter, cladding thickness of 0.0125 in., and with oxide fuel clad in stainless steel, the critical angle is 94°.

The temperature rises beneath such a bubble may be calculated using one of the three dimensional heat transfer codes which are available, TOSS or TRUMP (see the Appendix). The models used in these codes are the basic heat transfer equations detailed in Section 1.2 but arranged on a mesh that connects in three dimensions, making computer solution a necessity. Many of the difficulties involved in these calculations are connected with the spatial finite difference approximations used to represent conduction from one point in the mesh to another.

If the complete pin were blanketed by a bubble and the angle of coverage were 360°, then axial conduction would become important. Studies have shown that, for the above fuel pin, approximately an inch would have to be insulated by the bubble before failure of the fuel pin cladding could occur. For bubbles less than 1 in. in extension only moderate cladding temperature rises would result.

4.2.3 OTHER EFFECTS OF BUBBLES

The presence of bubbles has been noted as providing valuable nucleation sites (6) at which coolant boiling might start, if the heat removal from the system were inadequate.

In pure sodium, superheats of more than 500°F have been observed before boiling commenced due to overheating (7). If such superheats were possible in reactor accident conditions, then Table 4.2 shows that the time

TABLE 4.2

THE EFFECT OF SUPERHEATING ON CHANNEL VOIDING[a]

Superheat (°C)	Time to void channel completely (msec)	Mean excess channel pressure (atm)
500	30	10.40
200	59	3.04
100	95	1.22
50	151	0.49
20	293	0.13

[a] See Judd (6).

needed to void a channel completely would be much shorter and the excess channel pressures would be much greater. Thus, although accident conditions are to be avoided, if they can not be avoided then it would be preferable to have sodium boiling occur relatively quietly at little or no superheating.

The radiation present in the reactor core is expected to limit practical superheats to less than 200°C, but it has been estimated (6) that one bubble of radius 10^{-3} cm in every 100 cm³ of sodium would prevent superheats of more than 10 or 20°C. Absorbed gas appears to have little or no effect, but some entrained small bubbles are likely to maintain conditions at the lower end of Table 4.2. Fortunately, in a reactor system there will always be one or two small bubbles.

4.2.4 COLLAPSE OF SODIUM VAPOR BUBBLES

If a bubble of sodium vapor is surrounded by sodium liquid at the same pressure, thus providing a condensing medium, the bubble will then collapse. Initially the rate of collapse is slow and the rate of condensation keeps pace with it; however as the collapse rate increases the rate of condensation decreases as the heat transfer area is also decreasing. Thus the rate of condensation no longer matches the rate of collapse, so the internal bubble pressure will increase very rapidly indeed. The bubble then grows once more and collapses again and finally is fully condensed after several collapse and rebound cycles (6). The whole collapse and rebound cycle can occur in 10 msec and peak pressures of up to 8000 atm have been calculated for a 10-cm bubble. Such collapse pressures have been observed experimentally, although the energy possible in the pressure pulse is very limited. In the case cited it is about 4000 J. However this effect has not been observed in any reactor installation, although the phenomenon should be accounted for in accident analysis.

4.3 Criticality in Superprompt Accidents

Reactivity accidents have multiple safety features in the form of component-failure and flux-increase trips that initiate a shut-down of the reactor system before much has happened (beyond minor overheating of the fuel in the worst cases). This section therefore deals with those hypothetical accidents in which the reactivity addition is uncontrolled and in which eventual shut-down of the system is only achieved by a dispersion of the core.

Section 5.4 treats a discussion of design basis accidents; here we are simply concerned with the behavior of the system in the event of a complete loss of control and with the energy release from a superprompt critical system.

4.3.1 INITIATION

A reactivity addition accident that involves high addition rates attaining prompt criticality might arise from rapid voiding of the central assemblies of the LMFBR, an ejection of a control rod in some way, fuel slumping following local melting, or even from mechanical compaction of the fuel in a high-pressure fast reactor depressurization.

The improbability of these initiators is discussed in Section 5.4. It is sufficient to say that each could occur mechanistically, and therefore should be the subject of further study. These events could give reactivity addition rates in the range of from $ 5/sec to $ 200/sec (see Table 5.6).

4.3.2 REACTIVITY RATES OF ADDITION

Traditionally the worst rates of reactivity addition were calculated from a gravitational collapse of the core following gross core melting due to a loss of cooling. In many cases these were overemphasized by assuming that the center of the core had melted and collapsed before the top third of the core dropped onto the debris beneath. In this way, the worst possible addition rates were calculated.

In order to obtain some idea of how high rates are produced, consider a cylindrical core of height 123 cm and radius 100 cm which is presumed to melt and suddenly lose cohesion so that it slumps. The coolant is presumed to have evaporated so that the core has 30% coolant volume within which to slump. It would therefore fully compact, if it only slumped vertically downward, to a new height of 86 cm.

Equation (4.13) states the reactivity condition of the core before collapse, in which the square of the diffusion length L^2 and the dimensions of height H and radius R determine the leakage of the system.

$$k_{\text{eff}} = k_\infty \{1 + L^2[(2.405/R)^2 + (\pi/H)^2]\}^{-1} \tag{4.13}$$

When the core has collapsed to 70% of its height, assuming the diffusion length to be proportional to this height, Eq. (4.14) describes the new reactivity state k'_{eff} of the system.

$$k'_{eff} = k_\infty \{1 + L^2(0.7)^2[(2.405/R)^2 + (\pi/0.7H)^2]\}^{-1} \qquad (4.14)$$

If we assume that the system were originally subcritical due to the total core voiding, but that it just attains criticality on complete compaction, then $k'_{eff} = 1.0$. Substituting the values for R and H and assuming $L^2 = 150 \text{ cm}^2$, Eq. (4.14) gives $k_\infty = 1.1389$ and Eq. (4.13) then gives $k_{eff} = 0.9615$.

Thus during compaction the core gained reactivity due to a decrease in leakage of $1.00 - 0.9615 = 0.0385$ or approximately \$ 10. The gravitational collapse from 123 to 86 cm took 0.27 sec and thus the rate of reactivity addition was \$ 37/sec.

The rate could of course be much higher, if centrally situated fuel material could move further into the center where the material worth is higher. This particular illustrative example only accounts for leakage differences, and it assumes that L is only volume dependent.

Actually, the reactivity states of the changing configurations of fuel material would be more accurately calculated by using a two-dimensional diffusion theory code to calculate the effective multiplication several times during the collapse. This calculation would take into account the changing worth of the core fuel as well as leakage differences.

Section 5.4 outlines the practical accident analysis case in which the reactivity changes are actually a combination of voiding effects as well as fuel movement effects.

4.3.3 ENERGY RELEASE MODELS

The original mathematical model for the energy release following a reactivity addition was due to Bethe and Tait (8); although this model has been subsequently refined by a number of other workers (9, 10) and the calculational methods involving computers have improved, the basic principle of the model remains the same.

4.3.3.1 *Basic Calculational Procedure*

Figure 4.4 shows that the reactivity addition provides an increase in flux φ and, therefore, in energy E, which is contained only by negative feedbacks from the Doppler coefficient and eventually from a dispersion

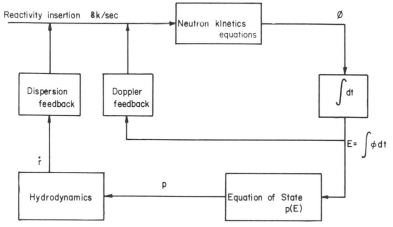

Reactivity insertion 8k/sec

Fig. 4.4. Block diagram for energy release analysis following a core disruptive accident.

of the core. The dispersion of the core is achieved because the high energy produced melts the fuel and the high internal pressures produced push the core aside into a subcritical configuration. The calculation terminates at this point because, in this model, there is no mechanism for recompacting the fuel material.

The equations which describe this model are the following.

$$k(t) = k_0 + k_{\text{input}}(t) + k_{\text{D}}(t) + k_{\text{disp}}(t) \tag{4.15}$$

$$d\varphi/dt = [k(t) - 1 - \beta]\,\varphi/l^* + S \tag{4.16}$$

$$E(r, t) = \int_0^t \varphi\, dt\, N(r) \tag{4.17}$$

$$k_{\text{D}}(t) = K_{\text{D}}(1 - \{E_0/[E_0 + E(t)]\}^{1/2}) \tag{4.18}$$

$$p(r, t) = \begin{cases} (\gamma - 1)\varrho[E(r, t) - Q^*] & \text{for } E > Q^* \tag{4.19} \\ 0 & \text{for } E \leq Q^* \tag{4.20} \end{cases}$$

$$d^2u(r, t)/dt^2 = -(1/\varrho)\,\text{grad}\,p(r, t) \tag{4.21}$$

$$k_{\text{disp}}(t) = \int u(r, t)\,\text{grad}\,D(r)\,d^3r \tag{4.22}$$

The effective multiplication (4.15) is a sum of the initial value, added reactivity diminished by Doppler k_{D} and dispersion feedbacks k_{disp}. In the neutron kinetic Eq. (4.16) the delayed neutron contribution S is assumed not to vary during the excursion. The power distribution $N(r)$ which defines the production of energy following the increase in flux (4.17) is assumed to remain constant in time.

Doppler feedback [Eq. (4.18)] is assumed to be dependent on the inverse of the square root of the temperature and therefore the energy of the fuel. The initial energy state of the system is denoted by E_0. The equation of state of the fuel [Eqs. (4.19) and (4.20)] defines the pressures p that are produced by the energy of the system E, although these pressures are not presumed to increase until the core is fully compacted and the fuel expands to fill the coolant volume. This initial expansion corresponds to an offset in energy Q^*. Once the pressures start to increase then a displacement of the fuel u occurs [Eq. (4.21)]. This equation is a normal hydrodynamic description of a fluid system. As in the equation of state, the density is assumed constant rather than taking account of a reduction of density due to the dispersion. This assumption of constant density means that the propagation of and reflection of pressure waves is ignored. It is a good assumption, as the fuel during dispersion does not expand much (less than 2%). The feedback reactivity due to the dispersion k_{disp} is calculated by knowing the reactivity change which would occur if a unit volume of the homogenized core at r were to be removed from the core. This value is called the reactivity worth function $D(r)$. Therefore, $u(r, t)$ grad $D(r)$ is the reactivity change involved in moving material from r to $r + u(r, t)$, and the feedback reactivity [Eq. (4.22)] is the volume integration of such changes.

Both the reactivity worth distribution and the power distribution, $D(r)$ and $N(r)$, respectively, are here assumed independent of time, ignoring the fact that these distributions will change as core materials move. Thus the reactivity changes are calculated from first-order perturbation theory. The assumptions are good for violent dispersions, as the volumetric expansions required to compensate for reactivity added above criticality may only be about 2 or 3%.

Using these equations, the total energy produced from this power burst may be calculated, as well as the residual pressures and core displacement. These values must then be interpreted in terms of the damage that could be done external to the core following the explosion. This subject is treated in Section 4.3.4; but first let us consider the sensitive parameters in the above set of equations and see what accuracy might be attached to the energy calculations.

4.3.3.2 *Sensitive Parameters*

The model above requires that the power distribution $N(r)$ and worth distribution $D(r)$ be provided as input functions. It is possible to use more exact diffusion equations to calculate these functions if a lattice model of the reactor system or core is used.

Thus Eq. (4.23) may be written for each mesh module, where the leakage $\nabla^2\varphi$ may be expanded in finite difference form depending upon the coordinate system chosen for the model.

$$l^* \, \partial\varphi/\partial t = [k_\infty(1 - \beta) - 1]\varphi + D\,\nabla^2\varphi \tag{4.23}$$

The most sensitive parameters in the calculation of the energy release are the following:

(a) *Equation of state.* Figure 4.5 shows typical energy release calculation results that vary by about a decade depending on which equation of state is used. Available data is limited and has been obtained in temperature and

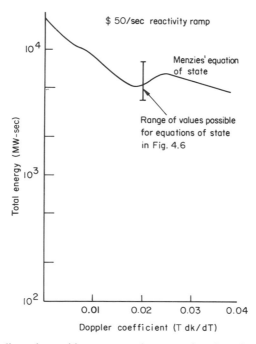

Fig. 4.5. Core disruptive accident energy release as a function of the equation of state used in the analysis.

pressure conditions well below those which are attained in these hypothetical accident conditions. Thus the extrapolation from available data to the region of interest is large. Not only does the available data vary, but the interpretations of this data also vary.

Figure 4.6 shows a number of equations of state in use at this present time for uranium oxide (*11a, b*). The most important version appears to

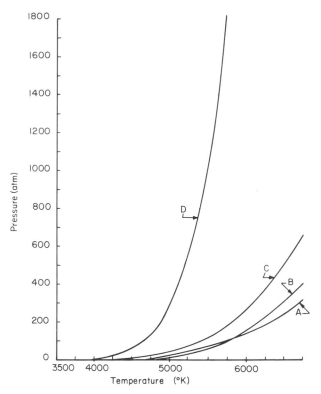

Fig. 4.6. A comparison of various approximations to the equation of state for uranium dioxide: A, Ackerman low temperature data extrapolation; B, Battelle Northwest Laboratory fit to available data; C, Menzies (*11a*); D, Ohse low temperature data extrapolation

be the fit to Menzies data (*11a*) from the Argonne National Laboratory:

$$p = 10^{-6} \exp[-4.34 \ln T - (76800/T) + 69.979] \qquad (4.24)$$

in which the temperature T is given in degrees Kelvin and the pressure p is defined in atmospheres.

In general, it would be preferable to have larger pressures for a given energy, so that as the energy increased the increased pressures would disperse the core more rapidly and thus lower the total energy release.

(b) *Reactivity addition rate.* Naturally, the faster the reactivity is added, the more the power rises and for a given Doppler feedback the energy release is generally higher. There are exceptions to this statement that depend on the fact that certain Doppler feedback values may be more effective at curtailing one power rise rate than another. This effect is detailed below.

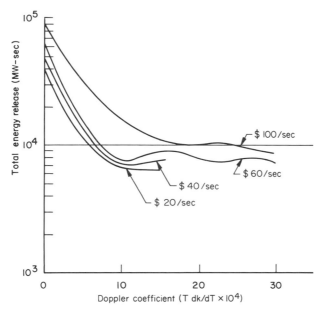

Fig. 4.7. Energy release as a function of the reactivity addition rate.

Figure 4.7 shows that the reactivity rate makes a large difference in the energy production when the energy release is high and only a small difference when the energy release is low.

(c) *Doppler coefficient.* Figure 4.8 illustrates the power transient for a given reactivity addition rate for very different Doppler coefficient values. In the case of the smaller Doppler coefficients, power values are higher because of the smaller feedback, and dispersion shuts the system down before the Doppler coefficient can really terminate the prompt power rise.

When a larger Doppler coefficient is involved, the power is generally held at lower levels and the power rise is turned over on Doppler feedback alone. However, a second power rise occurs as the reactivity addition rate still proceeds and dispersion later shuts the system down. Despite the fact that in the case of the larger Doppler coefficient power values are lower, the integrated energy release may be higher. From the cyclic occurrence of this paradoxical result the graph of energy release against Doppler coefficient obtains its oscillatory behavior.

Thus although, in general, a higher Doppler coefficient results in a lower energy release, for small changes in Doppler coefficient this may not be true. The Doppler coefficient makes a good deal of difference for small negative values, but after a value of approximately $-15 \cdot 10^{-4} \, T \, dk/dT$ very little

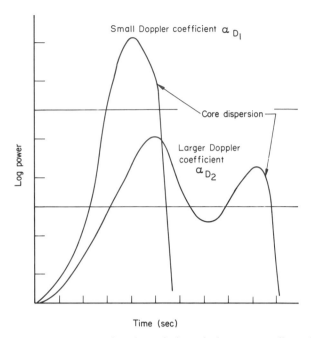

Fig. 4.8. Reactor power as a function of time during a core disruptive accident. The effect of high and low Doppler feedback coefficients is illustrated.

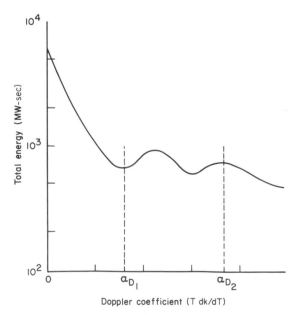

Fig. 4.9. Energy release as a function of the Doppler coefficient.

extra alleviation is obtained. Figure 4.9 presents typical results for an oxide fueled LMFBR.

A short delay time in the action of the Doppler feedback could affect the size of the energy release. Two such delays that have been postulated (*11c*), were the slowing-down time for fast prompt neutrons to attain the energies (about 1 keV) at which the Doppler feedback is generated, and the heat transfer time to attain the asymptotic temperature distribution between the UO_2 which is responsible for the Doppler feedback and the PuO_2 where most of the heat is generated.

Analysis however showed that, to be significant in a superprompt critical excursion resulting from a $ 100/sec addition rate, the delays should be greater than approximately 8 μsec. The first of the above delays in the slowing of neutrons to about 1 keV, is calculate to be 2 μsec and is therefore not significant. The second delay in PuO_2 to UO_2 heat transfer amounts to between 50 and 130 μsec for 40–60 μ grain sizes, and the delay should therefore be considered in evaluating the Doppler feedback. However the effect of such delays is not likely to be an overriding consideration in the energy release calculation.

(d) *Power distribution and reactivity worth distribution.* Figure 4.10 presents results as a function of the distributions of power and worth. In both cases the steeper the distribution in a single zone core, the smaller the energy release, although this effect is not so marked as the effect of the previous parameters.

(e) *The neutron lifetime.* The energy release varies with the value of the neutron lifetime but it is not a marked effect. It may be positive or negative.

In fact, the uncertainties associates with the above parameters in the calculation of the total energy release may be secondary to the uncertainties associated with the calculation of the proportion of this work that is available to do damage to the surrounding structure. This point is discussed in Section 4.3.4.

4.3.3.3 *Autocatalytic Effects*[†]

The above comments refer to an ideal single zone core, but present designs of sodium-cooled fast breeders comprise two zone cores in which the zones differ in enrichment; therefore, they have separate power and worth distributions. At the boundary between the zones there is a discontinuity in the

[†] See Nicholson (*12a*).

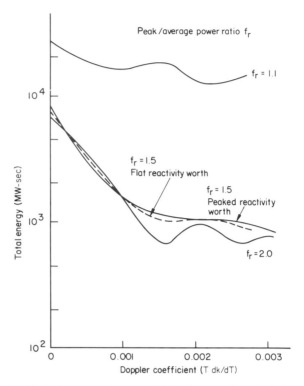

Fig. 4.10. The effect of power and worth distributions on the calculation of the energy release from a core disruption.

reactivity worth and in the power, so that fuel moving outward across the boundary between the zones experiences an increase in worth.

The discontinuity in power density produces an energy generation that is also discontinuous. A pressure discontinuity is produced that relieves itself as a compression wave inward and a rarefaction wave outward. Both waves can cause material to move inward near the interzone boundary.

Even in a homogeneous core there are also local areas in which flux and worth gradients do not coincide; this too can cause fuel material to implode rather than to disperse.

Calculations with the hydrodynamic code VENUS (*12b*), which includes a calculation of the reactivity effect of actual fuel movement rather than relying on a constant density material model for the expansion, have shown that the implosion effect is relatively unimportant. The inward implosion does not continue with ever increasing violence, but it rapidly explodes with a relatively small total energy production.

A calculation of the energy release, excluding consideration of the surface effects, may only be about 20% less than a calculation that includes all the surface or boundary effects.

4.3.3.4 *Available Codes*

The calculation of the energy release typified by Eqs. (4.15)–(4.22) is complicated by the need to solve spatial as well as temporal derivatives in a finite difference manner. Thus computer solutions are needed, and the available codes reflect improvements in the analysis over the past two decades (see also the Appendix).

(a) *AX-1 (13a)*. This is a one-dimensional spherical model with a linear equation of state. It accepted only a step of reactivity and had no Doppler feedback.

(b) *AX-TNT (13b)*. This development of AX-1 was also one-dimensional but it included a linear or Clausius–Clapeyron equation of state. It had partial Doppler feedback, and it accepted reactivity ramps as well as steps. It had a total of 320 mesh points and it used an additional routine to calculate the available work energy in terms of TNT equivalent.

(c) *WEAK EXPLOSION (14)*. Still a one-dimensional spherical model code, it included an equation of state of the form $p = B \exp[-A/(E + E_0)]$ and a Doppler feedback which depended on $E^{1/2}$. This code had no calculation of the work energy available for damage to structure.

(d) *MARS (15)*. This code is two dimensional (r, z). It allows in its modified form all types of equations of state, all forms of reactivity additions and allows the input of tabulated power distributions. Six core regions involving a total of 380 mesh points each can be used to map the core. Doppler feedback is included. A modified version of MARS calculated the available work as an isentropic expansion of fuel following termination of the transient by the dispersion of the core. However, modifications to the core model have allowed the available work to be calculated as a function of the amount of sodium present in the core into which energy is deposited by the molten fuel. This sodium is then assumed to undergo isentropic expansion.

(e) *VENUS (12b)*. This latest code is still under development at ANL. It combines a comprehensive hydrodynamic code that calculates the movement of fuel and structural material allowing a pointwise variation of density with time, with a point kinetic model of the kind used in MARS. The code is produced as a module in a much larger series of codes which starts with a dynamic module and ends with damage calculations within the vessel. It is the first to avoid the use of constant density, although the results do not differ greatly from MARS except in special cases.

4.3.3.5 Equation of State

The previous section has shown how the energy release calculation depends on the equation of state. Figure 4.5 shows several versions of extrapolations from three different sets of basic data.

To understand the importance of this equation in the calculation of total energy and, subsequently, the work energy, it is important to know how it fits into the phase diagram.

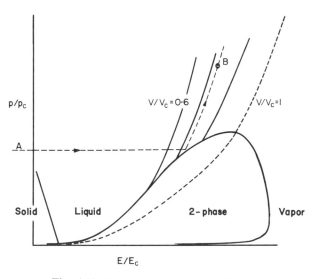

Fig. 4.11. Pressure versus energy diagram.

Figure 4.11 shows the two-phase diagram for fuel material in terms of the reduced variables pressure p/p_c, specific volume v/v_c, and energy E/E_c. These reduced variables are all unity at the critical point C.

The solid normal configuration fuel exists in a state illustrated by A. When the fuel is overheated it expands, melts, and further expands, all at constant pressure, until all the volume normally occupied by coolant has been filled. Then the fuel state will move along a constant volume line, now with increasing pressures, until some terminal state B is reached, defined by the fact that the reactor has shut down due to the dispersion. The constant volume line for the final configuration is usually about v/v_c of 0.7 as a core has approximately 30% coolant volume.

The path from A to B is the equation of state included in the energy release calculations above. It may pass through the two-phase region in certain cases.

There is very little data available at high temperatures and pressures, even for metal fuels let alone the oxides and carbides; thus the equation of state depends on extrapolations from low temperature data. An alternative method of obtaining the equation of state is to derive it from the generalized tables in terms of the reduced variables.

The law of corresponding states applies for materials with compressibility factors not very different from that of water (0.23). That for uranium oxide defined by Eq. (4.25) is 0.3.

$$\text{compressibility factor} = p_c v_c / R T_c \tag{4.25}$$

The path A to B in Fig. 4.11 is clearly a threshold function and indeed threshold equations of state have been used. Other versions in use are shown in Table 4.3.

TABLE 4.3

VERSIONS OF THE FUEL EQUATION OF STATE

Name	Equation[a]	Limitations
Threshold	$p(r, t) = (\gamma - 1)\varrho[E(r, t) - Q^+]$	$E \geq Q^+$
	$= 0$	$E < Q^+$
Linear	$p = \alpha\varrho + \beta\theta + \varepsilon$	
Clausius–Clapeyron	$p = \alpha \exp(-\beta/\theta)$	θ (keV)
Curve matches	$p = A \exp[-B/(E + E_0)]$	

[a] Symbols: p, pressure γ, ratio of specific heats
                 ϱ, density Q^+, threshold energy
                 θ, temperature α, β, ε, A, and B are constants
                 E, energy (initial value E_0)

In the accident analysis, the true core equation of state should not be that of the oxide or carbide fuel alone, but it should include allowance for the equation of state of the structural steel present as well as the remaining sodium in the core. Difficulty arises since the amount of sodium remaining in the core is generally unknown.

4.3.4 Work Production

The energy release calculations are performed in order to determine the amount of energy that is available to do damage to the structure surrounding the core. The ultimate objective is to be able to define the final position and distribution of the core debris, so that this debris can be maintained subcritical and adequately cooled. Thus we are interested in how much of the total energy release could be converted into damaging work energy.

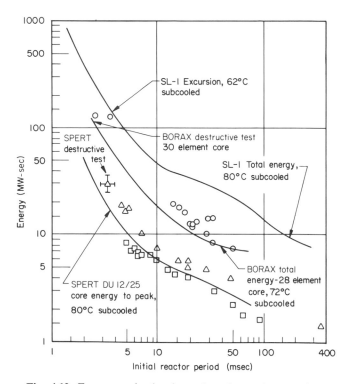

Fig. 4.12. Energy production in nuclear destructive tests (*16*).

Some idea of the magnitude of the damage which could be caused by nuclear explosions can be obtained from Fig. 4.12, which shows the energy generated in the destructive tests of the SPERT and BORAX reactors. Also shown is the calculated comparison of the SL-1 excursion (*16*). The energy ranges up to 200 MW-sec, which, from chemical explosion tests performed on scaled down reactor vessels, is equivalent to approximately 105 lb TNT.

After the power excursion is terminated by a small dispersion of the fuel material, the core will still be in the "constant volume" state assumed in the previous section. The coolant volume is occupied by fuel and structure debris. The center of the core is in a compressed liquid state under high pressures, and there may be some vaporized fuel in the core center while the periphery may be liquid or even solid. There is no sodium within the core, although there will be some sodium surrounding it.

The details of this homogeneous model of the disrupted and collapsed core depend on the severity of the energy release. From this state the core now expands and, in doing so, it does work. There are two basic expansion processes.

4.3.4.1 *Isentropic Expansion*

Figure 4.13 shows the state of the system in its initial state of compression at point A. It has an internal energy of α Btu/lb. If it now undergoes an isentropic expansion all the way to 1 atm, then it finally arrives at a state with an internal energy of γ Btu/lb.

Thus, in expanding, work has been done by the core on the surrounding material. From the first law of thermodynamics

$$\text{work done } dW = -dU = (\alpha - \gamma) \tag{4.26}$$

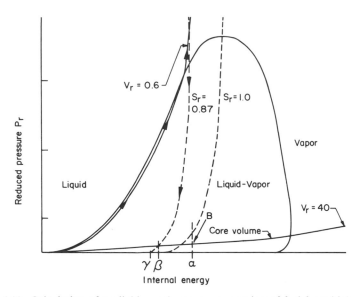

Fig. 4.13. Calculation of available work energy: expansion of fuel from high pressure condition during the core disruptive accident; V_r—reduced volume, S_r—reduced entropy.

This work is available to do damage to the structure around the core. The rest of the energy (γ Btu/lb) remains in the fuel material to be released on a somewhat slower basis to the surrounding sodium through heat conduction. This residual heat forms a large proportion of the total energy release and following transfer to the sodium, which may vaporize, further damage can be done.

The calculation of the exact isentropic expansion process can be performed if generalized tables are available. However, this is not so for uranium oxide.

Hicks and Menzies derived an approximate analytical method (9):

$$\text{work} = U_2 - U_1 - U_1 \ln(U_2/U_1) \tag{4.27}$$

The available work is the change in internal energy minus the unavailable work during the irreversible change from state 1 to state 2.

$$\text{work} = U_2 - U_1 - \text{unavailable work during}$$
$$\text{the energy generation phase} \tag{4.28}$$

$$= U_2 - U_1 - T_1(S_2 - S_1) \tag{4.29}$$

$$= U_2 - U_1 - T_1 \int dQ/T \tag{4.30}$$

Assuming that $dQ = c_v\, dT$ for a constant volume process, but that for a liquid c_v is approximately c_p and then using an average value of $\bar{c}_p$, Eq. (4.34) becomes:

$$\text{work} = U_2 - U_1 - \bar{c}_p T_1 \ln(T_2/T_1) \tag{4.31}$$

This expression is similar (17) to the Hicks and Menzies version in [Eq. (4.27)] but it does not assume that $\bar{c}_p T_1 = U_1$. Actually $U_1 = \bar{c}_p(T_1 - T_m)$ where T_m is the base temperature chosen as the fuel-melting temperature.

4.3.4.2 *Constant Internal Energy Expansion*

The compressed liquid could expand down a constant internal energy locus. Equations (4.27) and (4.31) give the work that *can* be done by the fuel, but if no work is done, then the expansion is at constant internal energy. In this case the expansion is down to the core volume rather than to 1 atm and if the sodium is added into the expanding system, assuming that rapid heat transfer has brought the fuel and sodium into thermal equilibrium, then very large residual pressures can exist at point B.

The isentropic expansion process leads to a maximum available work energy which may be used for pessimistic shock damage calculations for the vessel walls. The constant internal energy expansion, on the other hand, leads to pessimistic subsequent quasi-steady-state pressures in the system. In fact, the hypothetical expansion would be somewhere between these two extremes.

The available work energy derived from an isentropic expansion is not always the same fraction of the energy above melting; on the contrary, for small energy releases it is a very small fraction, while for large releases the work available can approach half the total energy above melting. Traditionally this work energy has been employed in subsequent damage calculations by equating it to an equivalent amount of TNT explosive in order to compute damage to the vessel and the structure surrounding the core. Section 5.5 treats this problem of energy partition in some detail.

4.3.5 IMPROVEMENTS IN THE ANALYSIS

Since the original Bethe–Tait calculations (8), the calculation of the energy release has improved significantly in the following ways:

(a) Doppler feedback coefficients are now included; they provide considerable amelioration of the energy release.

(b) Reactivity changes are calculated from a worth function $D(r)$ rather than from a one-energy-group perturbation theory and in the latest codes the reactivity changes are continuously computed by a diffusion equation calculation.

(c) The power function is now included as a shaped function.

(d) Two-dimensional geometry is used instead of spherical geometry.

(e) Improved nonthreshold equations of state are now used.

(f) It is to some extent possible to investigate nonhomogeneous autocatalytic effects to see what aggravation of the accident might be caused.

It has been pointed out that the history of energy release calculations has been marked by a succession of advances in calculational techniques that generally tend to diminish the energy releases calculated and a succession of conjectural physical occurrences, such as an implosion of the core, which tend to give greater explosive values. Fortunately over the years, although the energy release has been a saw-toothed curve, the tendency has generally been downward. So that, although the fast reactors considered have grown larger, the energy releases calculated have not grown markedly. Present day explosive values are at a level where they can be contained in containment barriers which the system might possess for other reasons (e.g., the reactor vessel).

4.4 Local Failure Effects

Local failures are defined here as those which cause local damage within a fuel assembly. The subject deals with the failure of individual fuel pins and the possible propagation of this failure from pin to pin and the subsequent possible failure propagation from assembly to assembly.

Section 3.1.1 dealt with fuel failure criteria arising from conditions within a fuel pin at the moment of failure, in particular for the case of fuel undergoing a power transient. It also dealt with changes of fuel configuration within the pin during burn-up.

This section is concerned primarily with conditions following failure, especially with how neighboring pins might be affected.

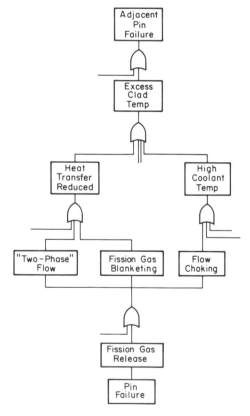

Fig. 4.14a. Fault tree for pin-to-pin failure propagation as a result of a release of fission gas from a failed pin.

4.4.1 Fuel Pin Failure

Figure 4.14 shows a fault tree for the undesirable event of a failure of a second fuel pin as a result of an original failure. It is conveniently divided into two cases: that in which only fission gases may emerge from the failed pin and that in which molten fuel is present and may emerge as well. These cases are very different because the emergence of molten fuel could cause sodium coolant vaporization and subsequent high pressures.

4.4.1.1 *High Burn-Up Pin Failure*

It is assumed now that no molten fuel is present. The problem is whether, following the cladding rupture, the release of fission gases might blanket

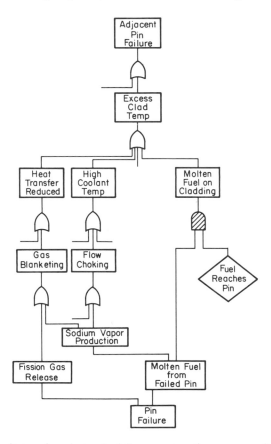

Fig. 4.14b. Fault tree for pin-to-pin failure propagation as a result of a release of molten fuel and fission gas from a failed pin.

and insulate an adjacent pin and cause a secondary failure. The results quoted below all refer to fuel pins of 0.2–0.25-in. diameter in a pitch-to-diameter ratio of about 1.25. The case is a very important one because the event of a failure of a high burn-up pin is a likely fault by the very statistical nature of the fuel and cladding fabrication process. Indeed plants have been designed to run with some failed fuel.

What experimental evidence there is suggests that no damage results subsequent to the first failure. The Soviet BR-5 was deliberately run at higher then design burn-ups for some considerable time and despite a significant number of failures, no report of any propagation mechanism was discovered (*18*).

Table 4.4 outlines results obtained in the USA for a 19-rod bundle (6 ft by 0.24 in. diameter with a 50-mil wire wrap on a 12-in. pitch) in which gas was released through different sized holes at different flow rates. Photographic records of the experiment, staged in water, showed that for large hole sizes, the outgoing gases could cause the channel flow to reverse but only for a very short time. Smaller hole sizes sustaining a longer outflow of gas caused no reversal and no coherent channel bubble but merely a stream of bubbles (*19, 20*).

TABLE 4.4

RELEASE OF FISSION GAS INTO COOLANT CHANNELS[a]

Hole size	Gas pressure, psia					Reference
	300	500	600	750	1000	
Rectangular hole (0.25 in. × 0.06 in.)	—	A gas bubble results which reverses flow. At 25 ft/sec the reversal is 2 in. and lasts 50 msec. At 17 ft/sec it is 4 in. for 100 msec.				*19, 20*
Circular hole (0.06 in. diameter)	—	No gas bubble and no reversal of flow experienced.				*19, 20*
Circular holes (diameters 0.005–0.01 in.)	Regions of opposite fuel pin experience smaller flow rates down to 5% of normal. Very small regions down to zero flow.		—	—		*21*

[a] See Carelli and Coffield (*19*), Hogaland (*20*), and reference (*21*).

Table 4.4 also outlines other work for a 7-pin bundle in water (*21*). Air was released through holes of varying sizes at various pressures into the coolant flowing with a coolant velocity of 20 ft/sec. The coolant pressure was 30 psi. No reversal was observed in the coolant flow, although measurements of flow against the opposite pin showed regions where the effective flow was reduced to as little as 5% and very small regions where the flow was zero.

Analytically, the consequences of fission-gas release within a coolant channel may be bracketed between large and small flow rates and their different consequences. First the rate of emission of gas may be calculated for a given rupture, depending on the pressure differential between the fuel-pin fission-gas plenum and the external pressure in the channel, if one knows the pressure drops along the path through the fuel by which the fission gases emerge (*19*).

Figure 4.15 shows a cross section of the fuel and the sintered and un-

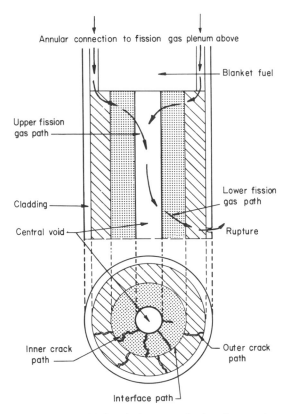

Fig. 4.15. Model for fission gas ejection flow paths.

sintered regions. Also shown is a postulated path for the flow of fission gas from the plenum through cracks in the outer region, through interconnected pores in the inner zone to the central void and a reverse path through the sintered and unsintered zones to the rupture of the cladding. Digital codes (see Appendix) exist to calculate flow rates through this tortuous path based on estimates of pressure drops that might be expected. The calculation takes into account the fact that the fission gas exits at sonic velocity, as long as the pressure ratio across the cladding rupture is greater than about 2 and thereafter the release is subsonic. Thus the gas outflow is calculated for different hole sizes, for different locations of the rupture, and for different assumptions of the condition of cracking in the fuel which determine the pressure drops through the fuel.

Fission gas outflows can last anything between a few milliseconds for large holes to even minutes for very small holes. Subsequent channel flows depend critically upon whether the flow rates are large or small as Table 4.4 shows; therefore, the analysis is separated into these cases.

(a) *Large gas flow rates* (Fig. 4.15). Based on a momentum balance, the flow rate in the channel can be shown to decrease rapidly when the fission gas enters, but it then recovers as the fission gas is swept out of the channel, until the flow regains its original value. This reducing and recovering flow transient can be input to a thermal assessment code to calculate the transient effect on the cladding temperature for neighboring pins. Even in the most pessimistic cases, the outflow of gas is too rapid to cause more than a moderate rise in neighboring cladding temperatures.

(b) *Moderate gas flow rates*. In these cases no flow reversal is assumed but it is supposed that the flow rates are large enough to allow a jet of gas to impinge upon and blanket an opposite fuel pin. Other results show that this can occur for a certain critical range of rupture sizes: between 0.005 and 0.07 in. in diameter (*21*).

A three-dimensional code TOSS (see Appendix) used to calculate the heat transfer for a pin that is effectively insulated for an angle θ of its circumference shows that the cladding temperature beneath the point of insulation rises to a level which depends on θ (Fig. 4.16a). As Section 4.2.2.2 showed, this level is much higher for coverages of larger than about 90°.

The cladding will attain failure temperature (*5b*) if the insulating or heating impingement of fission gas remains *in situ* for a long enough time (t_θ) which varies with the angle of coverage $\theta°$. The value of t_θ will depend on the characteristics of the jet and its possible deflection by the stream (*19*).

The following empirical equation can be used to represent the jet deflec-

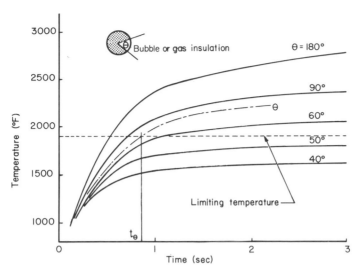

Fig. 4.16a. Cladding temperature resulting from vapor blanketing of the fuel pin as a function of the time and the angular extent of blanketing.

tion (velocity in feet per second and dimensions in inches):

$$x/d_0 = (2.02\ \varrho_g V_g^2/\varrho_c V_c^2)\ \log[1 + (0.049y/d_0)] \tag{4.31a}$$

where x is the penetration into the mainstream, y is the downstream deflection in a channel of diameter d_0, and ϱ_g, V_g, ϱ_c, V_c are the density and velocity of the gas and coolant, respectively. The deflection y varies with time as the gas velocity decreases through the rupture and t_θ is the total time that a point on the adjacent cladding has been blanketed by the jet of gas.

The jet also disperses in the flow according to the equation

$$\text{jet width } h = 2.25d_0 + 0.22x \tag{4.31b}$$

and this jet width will define θ, although of course the jet will also tend to smear around the adjacent pin thus increasing the effective blanketing coverage.

Thus failure can only occur by this mechanism if the cladding rupture is such as to produce a blanketing of the opposite pin of $\theta°$ for a time long enough to achieve failure temperatures t_θ. This demands that the rupture size be large enough to give a reasonable coverage, since smaller holes will give a smaller gas flow rate and therefore the jet may be deflected too greatly by the stream. On the other hand the rupture size must be small enough to sustain a gas jet for a time greater than t_θ. Thus there is a critical range of rupture sizes which might cause an adjacent failure.

Recent calculations (*19*) report that this critical range is between 10^{-4} and 10^{-5} in.^{2} in area.

Even within this critical range, there is yet no hint of the possibility of a continuing failure, for if a second failure is produced, it will still be directed back toward the original ruptured pin. In order to produce a chain of failures the second pin would have to also fail with a rupture size within this same critical range and at the same time be directed toward a third pin. This does not seem likely.

(c) *Low gas flow rates.* If the gas flow rates are very low then they will be sustained from small holes or locations distant from the fuel fission gas plenum for several minutes. The conditions in the channel are now two-phase and the analysis of Section 4.2.2.1 can be used to compute the effect on neighboring pins. This effect has been seen to be small for even void fractions approaching 50%.

Thus by separating the analysis into three distinct flow regimes, it has been possible to calculate temperature changes in the cladding of the neighboring fuel pins to determine whether any subsequent failure is caused that might lead to propagation of the original failure. The three analytical regimes are not strictly defined, they overlap although it appears that with internal fission pressures of 800–1000 psia the gas cannot cause a flow reversal for hole sizes greater than 0.06 in. in diameter at these flow rates (*20*).

The results lead to the following conclusions:

(a) For single failures the most critical size of rupture appears to be in the neighborhood of 0.01 in. in diameter (10^{-4}–10^{-5} in.^{2} in area) when a jet can be sustained to blanket the opposite fuel pin.

(b) For single failures, the most critical rupture location is at the middle of the fuel pin; ruptures nearer the plenum release the gas too rapidly and ruptures further away do not provide a very large flow volume rate.

(c) Even at these most critical conditions, temperature rises in the cladding of the adjacent pins are limited to very local positions, so that if a failure occurred in an adjacent pin, then the failure would be directed back toward the original failure. Thus it is most unlikely that this failure could propagate. Experience to date certainly substantiates this position.

Although once a problematic condition, it appears that experience and analysis have shown that a failure which releases only fission gas does not propagate. Because of this, experimentation on this condition is now being brought to a close.

4.4.1.2 *High Burn-Up Pin with Molten Fuel Present*

If molten fuel existed in the failed pin, then an additional phenomenon would be possible: Along with the ejection of gas, molten fuel might emerge into the coolant channel and there react violently with the sodium, causing it to vaporize. The violence of this reaction would depend on: the fragmentation of the fuel and the speed at which it could transfer its heat into the relatively colder sodium; how much sodium the fuel sees; and how it might be blanketed by the associated fission gas.

The analysis of the ejection of molten fuel (*19*) uses a model of the fuel which closely follows that for fission-gas ejection (Section 4.4.1.1). Figure 4.16b shows the assumed configuration of molten fuel, in, say, an over-enriched pin, with its exit path and the fission-gas-pressure driving force. The momentum balance for the molten fuel is given:

$$d(Mv)/dt = (P_1 - P_2 - \delta P)A \qquad (4.32)$$

In the molten fuel P_1 is the vapor pressure initially but it rapidly reduces to the fission-gas pressure after the ejection of only a very small amount of fuel through a fuel crack and the cladding rupture caused by contact

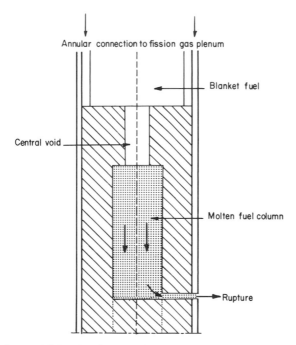

Fig. 4.16b. Model for the ejection of molten fuel from a ruptured fuel pin.

with molten fuel. The fission gas commences in a high burn-up pin at about 400–1000 psia pressure and then the gas loses pressure in the plenum as molten fuel leaves the pin creating a greater volume for the gas behind it.

The pressure in the channel P_2 does not remain constant because the molten fuel causes immediate coolant voiding as heat is transferred to the coolant and the pressure rises within milliseconds (see Section 4.4.1.4).

The pressure drop δP between the fission-gas plenum and the rupture site is a combination of many pressure losses on the tortuous path: molten fuel expansion losses, the fission-gas pressure drops, but no molten fuel pressure losses since these are likely to be small

$$\delta P = 1.2K\varrho v^2/2g + Av(26.05 + 20.5 + 0.21)$$
$$= \text{expansion losses} + \text{fission gas pressure drops} \qquad (4.33)$$

Equation 4.33 shows typical values calculated for crack frictional losses for the upper flowpath, the inner crack and the outer crack, and inner interface paths. These values are particular to a given fuel condition and would have to be calculated for each assumed differing configuration.

The mass balance for the molten fuel is determined by how much leaves

$$M = M_0 - \varrho A \int_0^t v\, dt \qquad (4.34)$$

Using the ideal gas law to calculate P_1 together with Eqs. (4.32)–(4.34), the velocity and mass of the ejected fuel can be calculated together with the reduction of gas pressure inside the fuel.

For a constant channel pressure P_2, for ruptures of 10^{-4} in.^{2} or less, a continuous jet of molten fuel escapes at velocities as high as 60 ft/sec for as long as 10 sec. For larger hole sizes, of the order of 10^{-2} in.^{2}, the velocity may pass through a maximum of about 10–30 ft/sec and the whole process could be over in 10–100 msec.

However, the situation changes radically if the fuel-to-sodium heat transfer is sufficient to cause high channel pressures, for as soon as $P_2 + \delta P$ attains the value of P_1, than the ejection ceases until the channel pressures are reduced and then the ejection recommences. Thus the ejection may be cyclic and will last longer than calculated for a constant channel pressure.

4.4.1.3 *Condition of Molten Fuel on Release*

It is important to know whether the heat transfer rates to the sodium are high and can cause rapid sodium vaporization and high channel pressures. These high heat transfer rates can arise if the fuel is well dispersed in fine particles.

TABLE 4.5

FUEL FRAGMENTATION ON CONTACT WITH SODIUM[a]

Fuel particle size (mm)	Percentage of total[b]
>0.3	81.0
0.22–0.30	12.0
0.15–0.22	4.5
0.08–0.15	1.5
<0.08	1.0

[a] See Amblard *et. al.* (*22*).
[b] From experiment with 1 gm of UO_2 in contact with 10 gm of sodium.

Experiments in Grenoble by the CEA and at Argonne National Laboratory (*22, 23a*) have measured sphere sizes and fragmentation following the injection of molten UO_2 into sodium. Table 4.5 shows the Grenoble data which agrees well with the ANL data.

Sphere sizes of 0.05 and 0.003-in. diameter were measured, and correlated to Weber numbers of 10–20 fairly well for the conditions of the experiment where the Weber number is defined in Eq. (4.35).

$$N_W = D\varrho v^2/\sigma g \qquad (4.35)$$

Assuming a Weber number of 20 for a maximum sized particle, in reactor flow rates of 25 ft/sec then the maximum size of particle is 0.009 in. in diameter.

Using the heat transfer correlation for small particles given in Eq. (4.36) or (4.37), the heat transfer rates possible may be calculated for particles of this size

$$N_u = k(P_r R_e)^{1/2} \qquad (4.36)$$

$$N_u = 2 + 0.39(P_r R_e)^{1/2} \qquad (4.37)$$

Thus heat transfer rates between 24,000 and 210,000 Btu/hr-ft²-°F are to be expected.

As the fuel emerges from the ruptured pin, the path can be calculated by assuming a drag coefficient for the given particle size and Reynolds number of interest. Particles of 0.009-in. diameter at 32 ft/sec excess velocity above that of the sodium stream are decelerated in a few milliseconds, and they move upward with the sodium and at the same time the heat transfer rate decreases as the relative velocity decreases. The relative velocity can decrease to 2 ft/sec in 1 msec.

Suppose however that the fuel manages to strike the opposite pin as a jet. Thermal calculations (23b) in which the jet effectively heats up the adjacent cladding show that the fuel pin is likely to fail within 10 msec depending on the angle of contact of the jet with the adjacent pin. These calculations assume that the cooling during transit across the coolant subchannel is small and it assumes a fairly small fuel viscosity (1 cP). If the viscosity is higher, then the Reynolds number and heat transfer rates would decrease and the situation would not be so severe. On the other hand the channel may be rapidly voided, so that the jet exists within the voided channel.

The foregoing considerations of fuel behavior are small sections of a larger picture which relate to each other. The next section considers the voiding of the coolant within the subchannel and then all these separate considerations are put together into a single description of possible pin-to-pin failure propagation modes. This final pieced-together picture is of course very dependent on the starting conditions and on the case being studied. We are here considering fairly typical LMFBR conditions for the first generation of power plants and therefore the final illustration should be generally applicable.

4.4.1.4 *Sodium Voiding*

Following fuel ejection sodium vaporization may occur due to the high heat transfer rates. The heat input into the channel may be calculated by using the previously calculated ejection rates with a calculated heat transfer rate, so that the final heat input is effectively of the form:

$$q/q_0 = [1 - \exp(-t/t_s)](P_1 - P_2 - \delta P)H \qquad (4.38)$$

The first term in this equation accounts for the delay time in attaining full mass flow. This heat input is used in conjunction with the voiding model described in Section 1.3 (24) to account for growth of the bubble and the rise of channel pressure (up to 1000 psia). In Eq. (4.38), t_s is the time constant for the fuel mass flow rate. Figure 4.17a shows the typical results which are produced. The size of the bubble is very small when the maximum pressure is attained; so condensation of the bubble does not have much effect on the maximum pressure value. However the maximum pressure is very dependent on the following parameters.

(a) The delay time expressed by $[1 - \exp(-t/t_s)]$ which represents that rate of ejection of molten fuel.

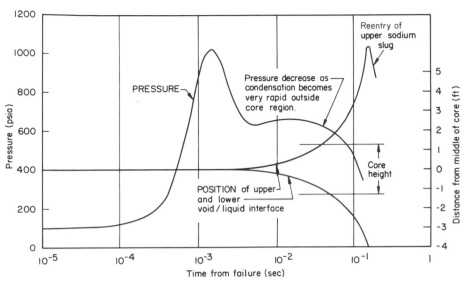

Fig. 4.17a. Growth of pressure and void length resulting from the interaction between molten fuel and sodium in a single subchannel (*24*).

(b) The heat transfer rate and therefore the fuel fragmentation represented by H in Eq. (4.38).

(c) The efficiency of the heat transfer process, which is a measure of the fuel fragmentation or possible vapor blanketing of the fuel during its ejection or the number of subchannels to which the voiding extends.

Most of the calculations performed to date assume 100% efficiency in transferring heat from the molten fuel to the sodium. However the amount of work energy which is derived from this heat energy is dependent on the model used to describe the voiding.

The ratio of work energy to original molten fuel thermal energy is defined as a reaction efficiency, and theoretical models for the molten fuel–sodium interaction may be checked for adequacy by comparing calculated efficiencies with values obtained from experiment.

Figure 4.17b shows the transient that resulted when uranium dioxide fueled pins were subjected to a high power transient in the TREAT facility piston autoclave. Sodium temperatures and pressure pulses were measured and correlated with the amount of fuel that was ejected from the fuel pins. The work energy was calculated from the motion of the piston as shown in the figure (*23a*).

The surprising result was a reaction efficiency of between 0.0015 and 0.018% and it seemed that the bond gas might be producing the pressure

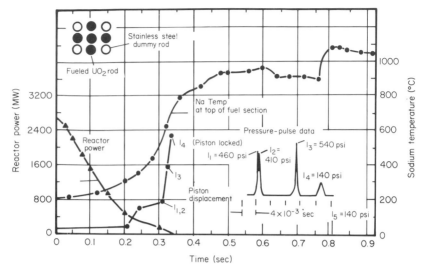

Fig. 4.17b. Behavior of UO_2 fueled rods during a high power transient in the TREAT piston autoclave. The reactor power is shown following the rise to maximum value; the rise in sodium temperature and piston movement up to its locked position are also shown. The inset shows the details of pressure pulses which occurred at times I_1, I_2, I_3, I_4, and I_5. They could have been related to the individual failures of the five fueled rods (*23a*).

pulses rather than their being the result of the fuel–sodium interaction. Subsequent experiments with evacuated fuel pins showed that the gas might be blanketing the interaction as the reaction efficiency rose to 0.15%. However the efficiency was still very low.

In laboratory experiments in which molten fuel (UO_2) was dropped into sodium, efficiencies of less than 0.01% have resulted although efficiencies of between 0.4 and 1.0% have been achieved with water injected into molten salt. The model described in Section 1.3 (*24*) is a pessimistic one, in that it results in a reaction efficiency in the neighborhood of 1–5%.

The voiding calculations should be linked to the fuel ejection calculations to provide a more accurate estimation of the channel backpressures. The voiding of a subchannel is very rapid and the complete subchannel may be voided in as little as 10 msec. The radial extent of that voiding occurs with about 80% of the speed of the axial voiding. This voiding is so rapid that the assumption that a jet of molten fuel might affect the neighboring pins depends on whether that jet transfers any of its heat to the coolant. If it does, then the coolant is likely to vaporize and failure could propagate by vapor blanketing rather than by jet impingement.

Once the void extends into the upper blanket region, the condensation rate becomes very high and the pressure is reduced and the void–liquid interface begins to reenter in about 0.6 sec.

While the sodium is being voided from the channel, condensation takes place on all the pin surfaces and the film left on the pin surfaces increases. Eventually however when the film has built up, in about 0.15 sec, the reverse occurs and the film dries out, so that when the sodium reenters, it may arrive back in a channel in which the fuel cladding has failed following dry-out.

The exact course of the voiding of a subchannel depends on the assumption of the fuel behavior within that subchannel. It is possible to postulate fuel failure while the liquid is out of the channel, such that when the sodium tries to reenter, it is prevented from doing so by the high heat ratings present in the molten fuel. Thus the void increases again, starting a chugging motion that has been postulated as an extreme form of heat removal. It has also been suggested that some of the fuel material which is ejected into the channel may be carried out of the channel by the vapor–liquid interface in the form of solid frozen particles. Such a condition, if it occurred widely, could cause a reactivity decrease.

4.4.2 Pin-to-Pin Failure Propagation

The previous sections have outlined small parts of the whole picture: fuel failure, molten fuel jet impingement on adjacent pins, fuel fragmentation, fuel velocities in channel, and voiding mechanics. It is now important to try to draw these pieces into a whole description of the sequence of events. The consequences of fuel failure should be determined in sufficient detail to establish what protection can be provided and what probability there is of a propagation of the failure.

Section 4.4.1.1 has shown that the propagation of failure due to fission-gas blanketing alone is unlikely and only in some certain circumstances could a secondary failure be caused. In this case for primary ruptures in the region of 10^{-4} in.^{2} area, a secondary rupture could be formed immediately opposite the primary one across the subchannel. The only place a third rupture could be formed would be back on the original pin. This A to B and B to A sequence is unlikely to spread the damage across the sub-assembly, especially since each rupture size must be that critical size to just give rise to the necessary conditions for continuing the process of failure.

Section 4.4.1.3 showed that molten fuel could eject out of a pin which already contained fuel and cause a jet failure on the next pin. However, again the failure sequence would be of the A to B and B to A type, which is unlikely to provide a tertiary rupture.

Section 4.4.1.4 has however discussed the voiding of the subassembly and the mechanism for failure throughout the subassembly is provided. Table 4.6 suggests a failure sequence using the information from the previous sections (*24a*). (Essentially, it provides more detail for the earlier portions of Table 5.8, which describes the same sequence.)

TABLE 4.6

OVERENRICHED FUEL PIN FAILURE SEQUENCE[a]

Time (msec)	Conditions
0	Overenriched hot pin ruptures as molten fuel contacts cladding during minor reactivity transient
0.001	Subchannel voided around failed pin, pressure about 1000 psia
0.008	Whole assembly voided, pressure about 150 psia
(0.010)	(Failure of adjacent cladding due to molten fuel jet impingement)
0.025	Cladding failure on the adjacent enriched peak pin
0.035	Cladding failure on all enriched peak pins in assembly
0.08	Pins adjacent to original failure melt (about four or five of them)
0.15	Film on voided channels at maximum thickness following condensation
0.20	Film dry-out following reduction in thickness.
0.30	Molten fuel ejection ends following an intermittent ejection
0.60	*Sodium reentry* into voided channels (vapor explosion?)
3.5[b]	Enriched pins molten and start to slump in contact with assembly can
5.0	Normally enriched pins molten and start to slump
15.0–20.0	Assembly duct experiences heat fluxes up to 2×10^6 Btu/ft^2 hr

[a] See Graham and Versteeg (*23b*).
[b] Cannot occur if the reentering flow reestablishes itself and is not blocked.

The overenriched hot pin is presumed to fail as molten fuel contacts the cladding due to some minor transient and the molten fuel is ejected. The subchannel voids rapidly in 1 msec with immediate pressures of 1000 psia. Then the void spreads more slowly across the subassembly, so that the whole subassembly is voided in about 8 msec and the pressures have been reduced to 150 psia. At about this time the molten fuel might also, by jet impingement on the adjacent cladding, have caused a secondary failure.

Due to the voiding, the cladding will fail on the adjacent enriched peak pins in about 25 msec and on all pins in the assembly in about 35 msec. However this is not significant since little molten fuel is present. Then the pins nearest the failure begin to melt and molten fuel may appear from at most 4 or 5 near pins in about 80 msec.

As the void is growing, it is condensing on pins above the failure, and the film on these pins is growing and heating up those components. Later, however, the process reverses and the film dries out in about 200 msec. By 300 msec, the entire fuel ejection process is over from the primary failed pin as well as those near it that melted. Then the sodium vapor–liquid interface reenters in about 600 msec.

At this point, several things could occur, and although detailed calculations might help to clarify this point, experimentation on fuel element failure propagation will be the only way to clarify the actual course of events. The following could occur:

(a) The sodium reentering could impinge upon the molten fuel which is in the channel and cause a sudden vapor explosion much more violent than the original vaporization. It is considered that evidence shows this to be unlikely.

(b) The flow could reestablish itself and normal flow conditions could maintain cooling of the subassembly, even though cladding has largely ruptured.

(c) Molten cladding could have blocked four or five subchannels and the condition changes to a treatment of a local blockage. In this case more than 8–10 subchannels should be blocked before further failure can occur and calculations (*25*) have shown that the blockage should be coherent. Even 1% seepage through the blockage could provide adequate cooling to avoid anything but a slow subsequent continuation of the damaging process.

During this voiding process, the reactivity feedback is small, limited to less that 10¢ for an entire voided subassembly. However, if the fuel melting results in gross slumping, then the reactivity changes are likely to be larger. These fuel movement reactivity changes could be of either sign, as can the voiding effects. Previous failures in both DFR and Fermi (*26, 27*) have been marked by negative changes of power, due to failure-induced reactivity feedbacks.

Neglecting the uncertainty of effects at this point, if the pins are now completely deprived of cooling, the enriched pins will be completely molten at the midpoint cross sections in about 3.5 sec, while the unenriched pins will reach the same state in 5.0 sec if we presume this assembly to be made up of a mixture of enriched and nonenriched fuel. Thus the assembly duct would begin to see slumped fuel in contact with it at about this time or shortly afterwards.

4.4.3 ASSEMBLY-TO-ASSEMBLY PROPAGATION

While the failed subassembly is being subjected to the sequence of events just described, the adjacent assembly can be expected to respond to two effects in addition to that of power changes caused by feedbacks from the failed assembly. First, it can be expected to feel some effect of the high pressures produced by successive voiding explosions in the failed subassembly, and finally it may be subjected to high heat fluxes due to failed and molten fuel against the assembly duct.

4.4.3.1 *Pressure Response*

In the failed assembly the pressure may attain 1000 psia locally which reduces to 150 psia when the whole subassembly is voided. Thus the pressure transient is roughly a very rapid rise to the peak followed by an exponential decay over the next 8–10 msec to the lower value.

If the failure is in the center of the subassembly, then explosion work (*28a*) has shown that the subassembly duct may only experience an attenuated value of a third of the peak value of pressure. Static calculations for the response of the assembly duct show that even in the irradiated state the duct can withstand several hundreds of psia without failure. However, if the original failure is close to the subassembly duct wall, then the original localized pressure of 1000 psia could cause failure in the corner of a hexagonal boundary (*28b*).

In any particular case, inertial calculations may be performed on the subassembly duct since the usual duct thickness is sufficient for the duct to have a natural damped vibrational mode with a period of about 1 msec. It will therefore not respond to acoustic waves generated in shorter times. Thus the duct essentially sees a steady pressure acting upon it even though this pressure may be localized. Any acoustic wave will have a very small invested energy (1 or 2 W-sec) because the relaxation time is so short (40 μsec for a central pin). The questions to be answered include:

(a) Does the duct wall fail for localized pressures and over what axial extent does it do so?

(b) If the duct fails, what pressure deformation does the adjacent duct wall experience and does this deformation cause boiling of the coolant in the second subassembly?

If boiling is caused in the second subassembly, then of course this constitutes propagation of the failure, since there is nothing now to stop the process of damage from repeating itself though on a slower time scale.

However, calculations to date appear to show that even with considerable deformation, the adjacent assembly docs not experience boiling, even when the outer coolant channels are completely constricted (*28b*).

4.4.3.2 *Response to High Heat Fluxes*

As the fuel slumps against the subassembly duct wall, heat fluxes reach several million Btu/hr-ft². The exact configuration of fuel against the wall does not matter because, even with small amounts, the fuel farthest away from the wall is soon at its vaporization temperature of 6500°F if no credit is taken for heat removal in the failed assembly.

A heat transfer calculation for this slumped fuel and the two adjacent subassembly walls and the neighboring subassembly may be calculated for a simple slab model to determine whether boiling (and thus an effective further failure) can be induced in the second subassembly. Figure 4.18 shows some typical results in which the adjacent subassembly is analyzed for its temperature distribution, a distribution due to an incoming heat flux from its neighbor. The COBRA code (*29, 30*) was used to compute the cross flows and the temperatures in each subchannel. In the particular case shown, the highest coolant temperatures occur in the shut-down case in which it is assumed that the reactivity feedback from the failed assembly has given a high flux signal and tripped the reactor. In this case a maximum temperature of 1660°F is reached, which shows that boiling may occur only if the system is at atmospheric pressure. However, primary systems are usually slightly pressurized and boiling would not occur. Not only does the slumped

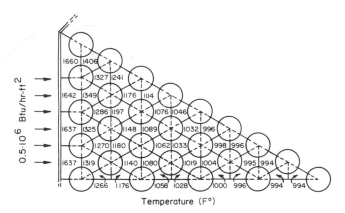

Fig. 4.18. Maximum coolant temperatures in an assembly following heat input from an external source for a particular illustrative case with the reactor shut down and the flow at pony motor flow rate of 10%.

fuel result in high heat fluxes to the adjacent can wall, but it also melts a certain amount of the can wall itself, attaining some sort of steady state in about 10–20 sec.

The final consequence of the subassembly failure depends on whether the reactor has been tripped or not due to reactivity feedbacks, whether or not the subassembly duct has been ruptured due to high pressure pulses, whether or not high heat fluxes have caused adjacent channel boiling, and what the final configuration of fuel in the failed subassembly is.

These answers depend on some experimental information that is still required: How does the film remaining after voiding grow and dry out, how does fragmented fuel behave in a voiding environment, how does fuel fragment, how do molten fuel pins slump, what is the response of the duct to pressure pulses within the assembly and locally against the duct wall, etc. It is expected that when these answers are forthcoming, surety will be obtained that failure cannot propagate beyond a single subassembly.

4.4.4 TOTAL CORE FAILURE

This section has so far treated failure arising from a single pin; however, it is possible that due to an overall reactivity accident or an overall flow failure, the whole core might almost simultaneously fail. The characteristics of the widespread failure would be very different. To obtain a widespread failure one must start with a hypothetical accident.

Section 5.4.4 treats suggested design bases and outlines the behavior following a loss of electrical power to the pumps associated with a failure to scram the reactor. The sequence of events in that case is as follows (remembering always that this is the supposed chronology of a hypothetical event):

(a) Widespread voiding due to either sodium boiling or fission gas release or both effects and consequent reactivity feedback.

(b) Prompt criticality resulting in widespread fuel failure and melting. A slight dispersion shuts the system down.

(c) Slumping of the core under gravity into a more reactive position.

(d) A second more violent superprompt criticality.

(e) Violent dispersion of the core.

The calculations leading to estimates of voiding times are the same as those previously described in Section 1.3. However the main difficulty in this analysis enters into the slumping calculation. Just how does molten fuel collapse? The time of collapse is all-important in defining the time scale in

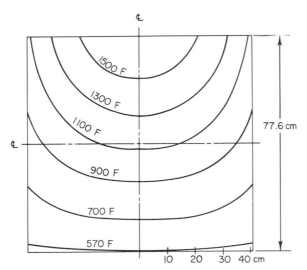

Fig. 4.19. Fuel temperature distribution in the core before the inception of boiling, at $t = 22$ sec. Fermi hazards report (*32a*).

which the slumping-induced reactivity feedback is added. This in turn defines the subsequent energy release.

Several fuel slumping models have been suggested in the past. One code, MELT II, allows the user to specify his own mode of fuel collapse for a single representative pin (*31*).

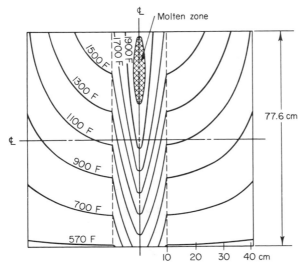

Fig. 4.20. Fuel temperature distribution in the core after the inception of boiling, at $t = 27$ sec. Fermi hazards report (*32a*).

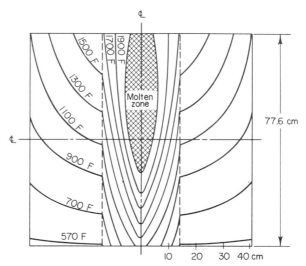

Fig. 4.21. Fuel temperature distribution in the core after the inception of boiling, at $t = 30$ sec. Fermi hazards report (*32a*).

Another method used in the Fermi hazards report calculated the molten patterns in a homogeneous core following a loss of cooling. These molten regions, shown in Figs. 4.19–4.24, were then assumed to slump under gravity (*32a*).

In several cases, EBR-II and Fermi, the hypothetical case of a gravita-

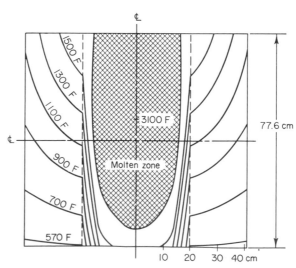

Fig. 4.22. Fuel temperature distribution in the core after the inception of boiling, at $t = 40$ sec. Fermi hazards report (*32a*).

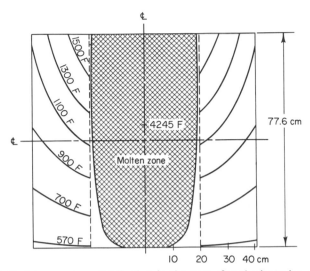

Fig. 4.23. Fuel temperature distribution in the core after the inception of boiling, at $t = 50$ sec. Fermi hazards report (*32a*).

tional slumping of the top third of the core onto a previously slumped and compacted bottom two-thirds was used as a design basis (*32b*).

Another suggested method is shown in Fig. 4.25 in which a representative fuel pin is modeled. The molten fuel distribution is assumed to run down as a collar in the equivalent channel of the pin and to successively grow as

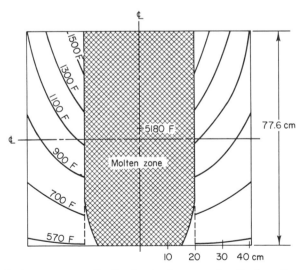

Fig. 4.24. Fuel temperature distribution in the core after the inception of boiling, at $t = 60$ sec. Fermi hazards report (*32a*).

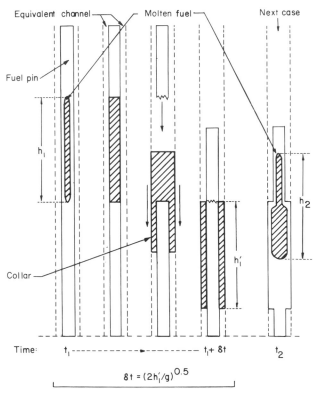

Fig. 4.25. Fuel slumping model.

more molten fuel is produced from the transient. The assumptions in this approach are that the core is voided and the void does not reenter while slumping is proceeding; there is no separated dripping of fuel; one rod is representative; the top of slumped collar is level with the top of the unmelted fuel; and there is apparently no hold-up of molten fuel by the surrounding unmolten material. The equations that define the configuration are those for conservation of mass and gravitational fall. The assumptions of this model are at least no worse than those of the others. However, they all suffer from one considerable defect: None of the core collapse models to date treat the combination of core voiding and fuel collapse and the interaction between the two phenomena, especially inasmuch as sodium liquid reentry is liable to have a compacting effect on the molten fuel. In some cases (*23b*) this interaction is not present, since the void exists for a long enough time for the fuel to slump while the liquid sodium is out of the core, but generally this may not be true.

Another hypothetical event which can lead mechanically to a core disruptive accident is an uncontrolled core reactivity increase. In this case the sequence is as follows:

(a) Power increase leading to fuel melting and eventual melting of the cladding from the high fuel heat fluxes.

(b) Molten fuel and fission gas ejection and fuel slumping into the coolant channel.

(c) Widespread voiding throughout the core (notice that in this case the voiding comes after rather than before the cladding failure).

(d) Widespread fuel slumping in combination with voiding mechanics.

This case is very different from the loss of flow combined with loss of scram, since in the previous case the voiding initiated the cladding failure, whereas here the fuel fails into a subchannel which is still filled with sodium. This latter event is likely to be more chaotic, although it is much less likely because of the design of the reactivity control system and the plant protective system.

4.4.4.1 *Experience*

Two fast reactor failures have resulted in extensive core damage.

In EBR-I, tests were being run involving reactivity ramps at low flows. As the power neared the safety limit, the order to shut the system down was misinterpreted and a slow shut-down was initiated instead of the scram system (*32b*). This occurred in November 1955. The sequence of events which followed was:

(a) Temperatures at the center of the core exceeded the NaK coolant boiling point.

(b) The boiling forced molten fuel material outward both above and below the core. (The fuel was metal uranium.)

(c) The channels were then blocked by freezing of the core material that then formed a cup into which further fuel material fell, including the tops of the pins.

(d) Fission products from the 0.1% burn-up fuel were released as gas bubbles, forming a froth or foam and thus a porous fuel mass when the fuel froze.

(e) Forty to fifty percent of the fuel melted before shut-down occurred.

The EBR-I core was very small and its 8-in. diameter core was hardly much bigger than today's assemblies and the fuel was highly enriched metallic uranium. The consequences of the failure are therefore hardly representative of possible events in a large distributed core. It is nevertheless representative of the speed of failure since all this took only a few seconds. A full and very readable account of the entire event is contained in the work of Thompson and Beckerley (*32b*).

The Enrico Fermi reactor suffered a melt-down of somewhat more than two assemblies in 1966 (*27*). This was directly the result of a blockage of the inlets to those assemblies and the accident was therefore a flow reduction. Fortunately the blockage occurred at assemblies in a low rated region of the core while the reactor was being brought up to power. Thus the power rating at the time of the accident was not high. The sequence of events which followed was:

(a) Presumably core voiding took place although this was not detected.

(b) Fuel cladding failed.

(c) Fuel movement within the failed assemblies gave a reactivity decrease. Later this was interpreted as being due to radial motion induced by the fact that the two assemblies melted close to each other, presumably because they were being cooled by the unaffected subassemblies on their other sides.

(d) This caused asymmetric fuel movement toward that side of the fuel subassembly.

(e) No fuel foaming was observed as had been observed in EBR-I, despite the fact that Fermi also had metal fuel.

More information is contained in Section 4.6.4 on the subject of this failure.

4.5 Sodium Fires

Sodium interacts with oxygen according to the reactions:

$$2Na + \tfrac{1}{2}O_2 \to Na_2O \qquad \delta H = -104 \quad \text{kcal/mole} \qquad (4.39)$$

$$2Na + O_2 \to Na_2O_2 \qquad \delta H = -20 \quad \text{kcal/mole} \qquad (4.40)$$

The interaction is characterized by low flames and dense white oxide smoke, which itself can create a visibility hazard for workers in the area. Once started, enough heat is liberated to maintain the reaction and gradually raise the temperature of the molten sodium.

4.5.1 POOL FIRES[†]

If sodium in a pool is exposed to oxygen, either because sodium is spilled in an air atmosphere or because what should have been an inert atmosphere was somehow contaminated or exchanged for air, then the possibility of a sodium fire is present.

However, the right conditions for that fire must be present. The sodium has an ignition temperature below which it will not ignite but will slowly oxidize. If the sodium surface is undisturbed, the ignition temperature is about 550°F, whereas this might fall to 400°F if the surface were disturbed. On the other hand some cases in which a sodium fire did not occur up to 800°F have been reported. There must be not less that 4% of oxygen present; otherwise there will only be considerable incandescence and a lot of smoke. This would occur all the way down to 0.1 vol% of oxygen. Humidity in the atmosphere is an important catalyst for a sodium fire.

If, however, a fire does start, the combustion rate is about 0.10–0.3 lb/min-ft² of surface. This rate would increase with the depth of the pool and decrease with the surface area of the pool. A fairly standard rate used in calculations is 5 lb/hr-ft².

Heat is generated at the surface and is dissipated from the sides and bottom of the sodium pool, so that there is a temperature gradient within the pool which might reach about 30°F/in. if it were quiescent. Above the surface of the sodium, there is a more severe gradient in "still" air of up to 500°F/in. Sodium fire modeling should take these possibilities into account.

When the sodium has burned completely, it liberates between 4100 and 4850 Btu/lb, although in most cases it does not seem to burn completely, and among the debris there is always unburned sodium.

4.5.1.1 *Mathematical Model*

In order to illustrate the main variables of interest in the sodium fire representation, the following model is worth discussion. The model comprises heat balance equations for the pool, flame, and the room, respectively, and mass balance equations for the sodium and the oxygen content of the room. The model is not a spatially distributed one; therefore the heat transfer from the room to ambient temperatures outside is necessarily crude. Nevertheless it does give reasonable results for large pools.

The pool loses heat to the vault walls at temperature T_v but gains heat from the flame at temperature T_f. Figure 4.26 shows the assumed configura-

---

[†] See Hines *et al.* (*33*) and Humphreys (*34*).

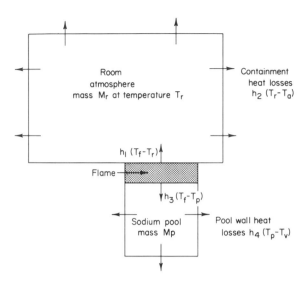

Fig. 4.26. Model for calculation of containment pressures following a sodium pool fire.

tion of the pool lying in a pit at the bottom of the room with a flame above it. Heat transfer rates are shown between the various components. Thus the pool heat balance equation is

$$M_p c_p \, \partial T_p / \partial t = h_3(T_f - T_p) - h_4(T_p - T_v) \tag{4.41}$$

The flame heat balance equation includes no heat capacity term, but it does include a heat production term based on the rate of burning of the sodium mass

$$-\delta H \, \partial M_{Na} / \partial t = h_1(T_f - T_r) - h_3(T_f - T_p) \tag{4.42}$$

The room heat balance includes heat received from the flame and that given up to the ambient temperatures T_a outside the walls.

$$M_r c_r \, \partial T_r / \partial t = -h_1(T_f - T_r) - h_2(T_r - T_a) \tag{4.43}$$

And finally the mass balance for oxygen depends directly on how much sodium is used; that is,

$$\partial M_{O_2} / \partial t = \lambda \, \partial M_{Na} / \partial t \tag{4.44}$$

The mass of sodium burned depends on the concentration of oxygen, and it is proportional to the square root of the absolute temperature as shown.

This is an experimental correlation.

$$\partial M_{\mathrm{Na}}/\partial t = -AcM_{\mathrm{O_2}}\,(T_{\mathrm{f}} + 273.2)^{1/2} \tag{4.45}$$

Finally the pressure in the room, according to the ideal gas law, is

$$p_{\mathrm{r}} = k(T_{\mathrm{r}} + 273.2) \tag{4.46}$$

In the preceding equations, δH is the heat of combustion (4850 Btu/lb), and c is a coefficient to make the units correct in the burning equation. The heat of combustion is based on an initial burning rate of 5 lb/hr-ft^2 and has a value of $0.17 \cdot 10^{-10}$ in cgs units. The heat transfer coefficients are all of the order of $10^{-4}\,A'$ cal/cm^2-sec-°C, where A' is the surface area of interest.

This model does not allow for the effect of a throat above the fire or for blanketing of the fire by the oxide the fire produces. It does not include time lags to account for nonimmediate mixing of the heat in the room or the pool, although this could be included by using a spatially distributed model. It naturally does not have enough nodes for more than illustrative accuracy. Nevertheless, very similar models are used in the sodium fire codes (see Appendix).

Figure 4.27 shows typical results that are obtained from such a model. After the fire is initiated by allowing contact between sodium and air, the temperature, and therefore the pressure in the room, rises rapidly to a peak where there is a balance between the heat input and that lost through the walls of the room. After this peak, due to lack of oxygen, the temperatures

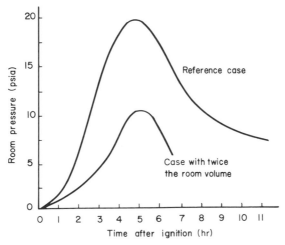

Fig. 4.27. Containment pressures due to a sodium pool fire.

slowly diminish. The time scale is long, the maximum pressure of 20 psia not being attained in this case for 5 hr. Also shown is a parametric case in which the room volume was doubled.

The pressures resulting from the sodium fire may in certain cases set the design pressures for the containment building (*34*) if nothing worse than this fire could be envisaged. The sodium and its combustion products may of course be slightly radioactive and the smoke is caustic.

The burning rate in the above model was based on a proportionality with the concentration of oxygen and the square root of the absolute temperature, as predicted by experimental work and theory. The square root accounts for the relative velocity of the sodium and oxygen molecules. If the reaction takes place in the pool, then the expression of Eq. (4.45) can be used, but if the reaction actually takes place in the flame, then the concentration of sodium molecules also ought to be included in the expression.

$$\text{burning rate} = c'AM_{O_2}M_{Na}(T_f + 273.2)^{1/2} \tag{4.47}$$

Some codes, instead of assuming a semiempirical burning rate of this kind, actually calculate the chemical balance at each point in time from the reaction between available sodium and oxygen molecules. This approach is more useful in spray calculations.

4.5.2 SPRAY FIRES

If the sodium is sprayed out into an atmosphere that contains oxygen, then again the possibility of a fire is present. The sodium spray may arise from the failure of a primary sodium pipe under pressure or a similar circumstance, although it is a little unlikely in an actual plant. Pipe leaks forming pools are more probable if a leak occurs.

When fine particles of sodium are sprayed into the air, the sodium may ignite at room temperature if it is in the form of a mist, although it may not ignite until the temperature is about 250°F if the sodium is in the form of droplets. The burning rate increases as the oxygen concentration is increased, up to 5%. It also increases if the moisture content of the atmosphere is increased, although there is no significant effect following an increase in the spraying pressure (*35*). Experiments have produced burning rates that are as rapid as the ejection rates themselves.

Heat from the spray fire is lost to the oxide particles, to the unburned sodium, to nitrogen in the air, and to the containment walls by radiation. Modeling of a spray fire must therefore take all these modes of heat transfer into account.

Due to the motion of the particle, the fire is less likely to be extinguished by oxide blanketing. Safety features that could be employed to protect against the consequences of spray fires include baffles to accumulate the spray volume thus reducing the heat transfer area and allowing the fire to be blanketed. A better safeguard is to provide inert gases only in proximity with sodium.

4.5.2.1 *Mathematical Model*

A model of a spray fire must take into account the heat balance of the burning particles, motion of those particles through the atmosphere, and a representation of a succession of particle waves moving into the atmosphere.

Figure 4.28 shows a representative particle that is assumed to exist in an atmosphere of radius $D(r)$. The heat balance for the spherical particle is:

$$\tfrac{4}{3}\pi r^3 c_p \varrho \, dT_1/dt = h_1(T_1 - T_2) \, 4\pi r^2 \tag{4.48}$$

Heat is lost to the flame of temperature T_2, which itself has a similar empirical heat production term as used in the pool fire representation. The flame also loses heat to the atmosphere and to the containment wall at

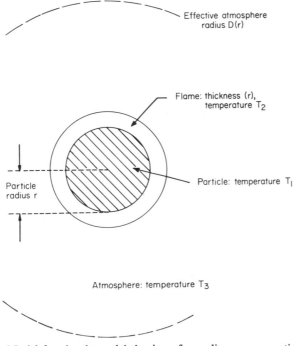

Fig. 4.28. Model for the thermal behavior of a sodium spray particle on fire.

temperature T_c

$$4\pi(r + d)^2 d\, \delta H\, M_{O_2} c(T_2 + 273.2)^{1/2} + 4\pi r^2 h_1(T_1 - T_2)$$
$$= h_2(T_2 - T_3)4\pi(r + d)^2 + R[(T_2 + 273.2)^4 - (T_c + 273.2)^4] \quad (4.49)$$

The atmosphere heat balance equation is:

$$\tfrac{4}{3}\pi(D^3 - r^3)c_a \varrho_a\, dT_3/dt = h_2(T_2 - T_3)4\pi(r + d)^2 \qquad (4.50)$$

As the sodium is burned up the radius of the particle naturally decreases according to

$$4\pi r^2 c_p\, dr/dt = M_{O_2} c(T_2 + 273.2)^{1/2} 4\pi r^2 \qquad (4.51)$$

The reduction of oxygen in the atmosphere is then proportional to the dimunition of sodium as burning proceeds.

$$dM_{O_2}/dt = \lambda 4\pi c_p \varrho r^2\, dr/dt \qquad (4.52)$$

The equations so far describe only the heat balance of the combustion itself. The constant R is a radiation constant equal to the emissivity multiplied by the Stefan–Boltzmann constant and λ is the proportional amount of oxygen needed for a given mass of sodium during the combustion.

The model does not allow for the production of the combustion oxides which absorb heat. It does not allow for the interaction between particles or for the inclusion of more particles than one simultaneously in the volume $D(r)$. It does not allow for the motion of the particles. However, all three effects can be included, the most difficult being the agglomeration processes involved in the interaction between the particles and between the particle and its surroundings (see Section 5.1.5).

The above equations, when solved, result in a temperature and a corresponding pressure rise (using the ideal gas law). The particle diminishes in size until the burning is complete and the injection of a succession of particles will eventually exhaust the oxygen in the atmospheric volume $D(r)$.

A simpler spray calculation assumes a homogeneous burning mixture of sodium within a total containment atmosphere with no allowance for fine structure details within the spray. This calculation results in large values for resulting pressure and temperature rises.

Experimental results for spray fires are less severe than those calculated. Figure 4.29 shows a comparison between experimental results for sodium ejected into an air atmosphere and calculations using instantaneous mixing, no heat losses, and infinitesimal particle sizes. The calculations are of course the theoretical maxima enveloping the experimental values.

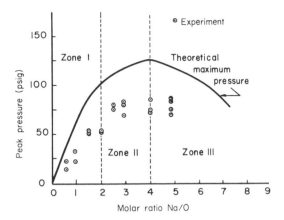

Fig. 4.29. Peak pressures due to a sodium fire as a function of the molar ratio of sodium to oxygen (*33, 34, 35*).

4.5.3 EXPLOSIVE EJECTION OF SODIUM

The only effective difference between explosive ejection and high pressure spray discharge is one of burning rate. Under given temperature conditions, the rate of reaction is a function which depends on the rate of sodium exposure, which in turn depends on the mass rate and particle size of the sodium discharge.

Experiments at Argonne National Laboratory (*35*) for explosive ejections of 400°F sodium in 3–10 msec have shown that the pressures produced are much less than theoretically possible (Fig. 4.29). It is clear that there are three zones of combustion:

(a) When only a small amount of sodium is present, the peroxide is formed an pressures are only moderate because the peroxide reaction has a lower heat release than the oxide reaction.

(b) When more sodium is present, more heat is produced as the oxide is produced, because the extra sodium reduces the peroxide.

(c) When even more sodium is present, it acts as a heat sink and it is very effective in reducing the overall pressure rise.

The rate of fallout of the reaction products is an important parameter in the pressure–time characteristic, since, in the experiments, over half the airborne reaction products had settled to the floor of the reaction vessel within one minute of the ejection.

4.6 Previous Experience

In any safety text a summary of previous accidents in similar systems is most important, because it is only by our experience that we learn. An excellent survey of all reactor accidents up to 1964 is contained in the work of Thompson and Beckerley, referred to in the Preface. The following section merely highlights those main operating experiences and occurrences which have taken place in fast reactors. No systematic classification is intended; rather it is hoped to show a general picture of the multitude of kinds of problem that arise and that have to be accounted for in a comprehensive safety evaluation.

4.6.1 SURVEY

No really large fast reactors have yet been built; the largest which are presently under construction are the Prototype Fast Reactor (PFR) in Britain, Phenix in France, and BN 350 and BN 600 in the USSR, all of which are above 550 MWt. In the USA, the Fast Flux Test Facility (FFTF) will include a 400 MWt reactor. Table 4.7 shows a listing of the main parameters of the most important reactors already built or in the planning stage.

Apart from CLEMENTINE which was cooled by mercury, all fast-reactor systems to date have been sodium- or NaK-cooled. There have been no gas-cooled versions.

Some of these systems have suffered operational teething problems before and after start-up, and others have experienced accidents and incidents during operation. These accidents have never resulted in any risk to the public, and only two have resulted in considerable material damage and delays. The following sections outline some of the major problems.

4.6.2 DOUNREAY FAST REACTOR (DFR)[†]

The DFR is a 60 MWt system cooled with down-flowing NaK. When built, the system used the present state-of-the-art which led to the use of small electromagnetic pumps and 6-in. diameter main loops. There are thus 24 loops of all welded construction doubly contained throughout. There are no valves or seals, and even the control rod drive is transmitted through the vessel by an electromagnetic clutch. The fuel element had a

---

[†] See Tatlock *et al.* (*36*).

TABLE 4.7

REACTOR CHARACTERISTICS AND SAFETY FEATURES

Reactor	Power (MWt)	Fuel loading (kgm)	Fuel	Heat transport and auxiliary cooling (ACS)	DBA	Containment
EBR-II	62.5	363	U	Pool	440 MW-sec	(1) Reinforced concrete (2) Steel
Enrico Fermi	300.0	2000	U	3 loop	1900 MW-sec	(1) Steel
SEFOR	20.0	1920	Pu–UO$_2$	1 loop, ACS	20 MW-sec	(1) Reinforced concrete (2) Steel
FFTF	400.0	3000	Pu–UO$_2$	3 loop	—	—
FARET	50.0	260	Pu–U	1 loop, ACS	60 MW-sec	(1) Reinforced concrete
PFR	600.0	6000	Pu–UO$_2$	Pool, cooler	Limited melt-down	(1) Primary tank (2) Containment building
SNR	730.0	5000	Pu–UO$_2$	3 loop, ACS?	2400 MW-sec (total)	(1) Reinforced concrete and dome (2) Steel shell
Rapsodie	20.0	300	Pu–UO$_2$	2 loop	(1) Disruption (2) Sodium fire	(1) Concrete blast shield (2) Steel shell
Phenix	560.0	3500	Pu–UO$_2$	Pool, cooler	(1) Disruption (2) Sodium fire	(1) Concrete blast shield (2) Steel shell

safety design with a niobium outer cladding and a vanadium inner cladding that was designed to fail first, in the event of any overheating, to allow any fuel debris to run down the center of the annular element. The fuel pins are restrained from bowing. All the control rods are worth less than $ 1; they are moving fuel assemblies inserted from beneath.

There is a thermal siphon capability for decay heat removal upon shutdown. The containment sphere is designed for a pressure range of from +18 psig to −4 psig. Final dispersal facilities are provided by niobium-plated cones inside and outside the reactor vessel, leading finally into a series of cylinders terminating in the bedrock. Figures 4.30 and 5.16 show reactor cross sections.

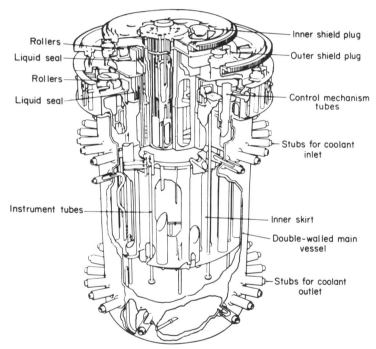

Fig. 4.30. Cross section of the Dounreay Fast Reactor (*36*).

4.6.2.1 *Downward Flow*

Downward flow was chosen for the DFR, because the top shield and mechanisms are kept cool at the reactor inlet and down-flow avoids levitation of the fuel subassemblies.

Against these reasons must be set the disadvantages that the coolant flow is in the opposite direction to natural circulation for the thermal siphon and is opposed to buoyancy forces. The cover gas must be pressurized and the instrumentation at the core outlet is very difficult because of its remoteness from the operating floor.

It should be added that a considerable number of problems have arisen due to the choice of down-flow; it is unlikely that such a choice would be made in any present or future system.

4.6.2.2 *Difficulties prior to Start-Up*[†]

Oxide contamination of the coolant was a major preoperational difficulty. Flow meters were erratic, 6 of the 12 rods could not be raised due to crudding on the rod mechanism which broke the surface of the sodium, and the cold trap circuits rapidly became clogged.

The coolant was dumped and cleaned, and the rod mechanisms were modified to avoid breaking the surface. Some of the purification circuit lines proved to be too small in diameter; they were eventually replaced by an external purification circuit. The dirty coolant was principally a quality assurance problem prior to the filling of the primary circuit and was cleared by draining and refilling with clean NaK.

Gas entrainment in the primary system was quickly observed. Expansion tank levels were erratic and, at low power, reactivity changes occurred due to the voidage in the core region. Some of the gas was being entrained through thermocouple and control rod guide tubes that dipped below the surface of the coolant. The entrainment problems in these areas were solved by drilling holes in the thermocouple guide tubes to equalize pressures inside and out and by installing "hats" on top of the control rod guide tubes, so that they could be supplied with an independent cover gas system. Both of these problems, coolant crudding and gas entrainment, were accentuated by the multiplicity of circuits in the system, since a single modification to a circuit had to be repeated 24 times.

4.6.2.3 *Difficulties after Start-Up*[†]

a. *Coolant contamination.* Coolant contamination persisted. It was eventually cleared by installing an external cold trap of much larger capacity, by adding plugging meters to detect impurity, and by additional pipework for balancing of cover gas volumes. Some of the now unused purity lines still remain blocked.

[†] See Smith (*4d*).

b. *Design changes.* Because of design changes, the original zero power physics calculations made on the ZEUS pile were invalidated. Therefore the reactor had to operate at 1 MW until it provided its own physics data to provide confidence for its high power operation.

c. *Reduced stability.* At high burn-up at 45 MW operation, less stability margin was apparent. This was apparently due to fuel expansion within cracks at the higher burn-ups, thus reducing the axial expansion feedbacks that were a negative contribution. This was no problem, but it pointed to the need for reactivity feedback measurements with adequate perturbations. The DFR oscillator had a 1¢ perturbation that was far too small for this investigation of feedback reactivity.

d. *Radial blanket overheating.* Due to maldistributed flow in the radial breeder at low flows, the blanket was severely damaged by stationary bubbles caught in buoyant down-flow conditions. Six hundred elements were removed, 200 only after cutting swollen sections. In fact, insufficient attention had been paid to the hydraulic design of the blanket sections, as these sections had low power ratings.

e. *Core fuel pin damage.* As entrained gas bubbles during start-up may adhere to unwetted fuel pins and then, due to this adherence enhanced by their buoyancy, stay long enough to blanket the pin as the power is raised, some fuel pins have been subjected to failure. This failure mechanism is troublesome in reactor operation but has been valuable in the sense that in no cases has there been any evidence that pin failure has spread or propagated (see Section 4.4.2) *(26)*.

f. *Primary circuit leak.* A small leak in the primary circuit caused a one-year operating delay in order to find and repair the leak. The location of the leak was very difficult to determine, because the reactor is a 24 loop system any one of which could have caused leaking sodium to appear at the point where it was detected in the leak jacket.

First the cover gas was pressurized without affecting the leak. Then 10 gm of gold tracer were introduced into the primary and the tracer's presence in the leak jacket subsequently confirmed the leak presence. Successive isolation of the leak location was made possible by adding helium to the cover gas and lowering the sodium level slowly to try to determine when the leakage stopped. This confirmed that it was in one of 24 bottom outlet pipes. Then, with a pressurized leak jacket and manual movement of the heat exchangers attached to each outlet pipe in turn, inward bubbling from the leak jacket was detected.

The failure was finally determined to be a fatigue failure of a bad and misaligned weld combined with thermal stressing due to cold NaK entering the hot outlet stream from a subsidiary purity line just upstream of the weld. Neither weld nor thermal stressing was sufficient to cause the failure alone. The purity line was removed and the weld remade to solve the problem.

4.6.3 Rapsodie[†]

Rapsodie is an originally 24 MWt system cooled by up-flowing sodium which has been uprated to 40 MWt in 1970. It is fueled with mixed plutonium-uranium oxide, stainless steel clad, and wire wrapped in 37-pin hexagons. The power rating is 340 W/cm³. Figure 4.31 gives the reactor cross section.

The vessel is hung for bottom entry, the loops are nonelevated but are all doubly contained. The two loops join before vessel entry and come in as a single saxophone pipe. Figure 4.31 shows a cross section of the reactor vessel and its nitrogen filled guard vessel. All the circuits have up to 4% natural circulation capability, and they are provided with EM pumps with double check valves.

Six control rods are driven in, and a digital computer is used for trip analysis since there are: 15 scrams on high flux, one each on γ activity in vaults, power-to-flow ratio, thermocouples at the reactor outlets, seismic disturbance, and manual trips. In addition there are 84 trips derived from each thermocouple of each subassembly outlet.

Failed fuel detection instrumentation takes the form of delayed neutron detectors in the primary sodium circuit and within the argon cover gas system.

The fuel handling equipment was a very complex machine handling more assemblies than one at one time. This was eventually changed for a simpler device, another instance in which simplicity improved reliability and safety. Emergency cooling is provided inside the double containment.

4.6.3.1 *Immediate Modifications*

The vessel plug had been designed to take several positions depending on a servosystem related to temperature and pressure measurements. This proved far too complex and moved when it was not required to do so. It was altered to a fixed position for simplicity.

The high-speed argon blowers were changed for conventional low-pressure

[†] See Denielou (*37*) and reference (*38*).

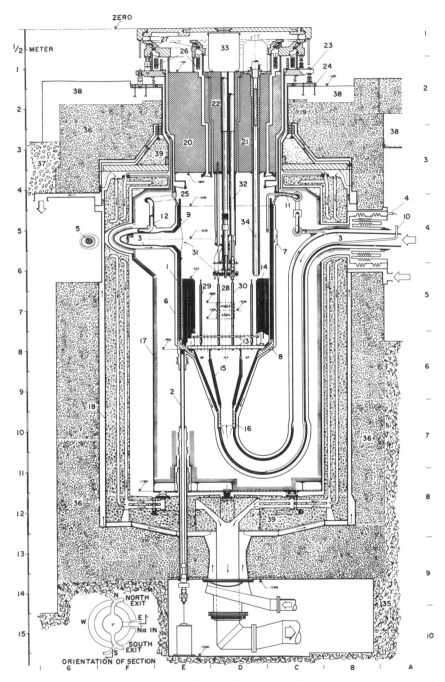

Fig. 4.31. Cross section of the Rapsodie Reactor (*39*).

IDENTIFICATION KEY FOR FIG. 4.31

Code no.	Position	Description
1	E5	Leaktight vessel
2	E6	Stereotopographical measurement position
3	B4–F4	Loop pipes (inlet and southern outlet)
4	A4	Nitrogen supply pipe for interspace II
5	G4	Southern purification loop
6	E5	Preheating jacket
7	C4	Nitrogen distribution annulus
8	C6	Preheating jacket keying
9	E4	Thermal shield
10	A4	Fill and drainage pipe
11	C4	Siphon breaker
12	E4	Safety injection pipe
13	D6	Support grid
14	D5	Neutron shield
15	D6	Diffuser
16	D7	Annular diaphragm
17	F6	Guard vessel
18	G7	Outer boundary
19	C2	Thermal deck
20	E3	Large rotating plug
21	D3	Small rotating plug
22	D2	Control plug
23	B1	Support flange for rotating plugs and leaktight vessel
24	B2	Support plate
25	E3	Anti-sodium barrier
26	E1	Liquid joint
27	E1	Lip joint
28	D5	Fuel assembly
29	E5	Breeder assembly
30	D5	Control rod
31	E4	Core cover
32	D3	Control rod guide tube
33	D1	Drive mechanism housing
34	D4	Handling guide tube
35		Ordinary concrete ($d = 2.3$)
36		Normal borated concrete ($d = 2.8$)
37		Heavy baryted concrete ($d = 3.5$)
38		Metallic insulation concrete ($d = 5.4$)
39		Rare earth concrete ($d = 2.4$)

Fluids used: cover gas—argon; main coolant—sodium; other interspace and preheating fluids—argon and nitrogen.

blowers with oil bearings. The more advanced components had previously been sticking.

Some of the previously inerted vaults below the operating floor were changed to air for convenience, although nitrogen was still retained within the reactor vault and around the primary system.

The air conditioning system within the containment began giving rise to a water collection problem that had originally not been recognized as a possibility. This problem was solved before start-up with the sodium-cooled reactor.

4.6.3.2 *Experience*

Rapsodie has operated with great reliability. For its first two years it had 86% availability with a 61% load factor. There were only 21 scrams and 30 setbacks in power. The reactor had as many as 43 in-pile experiments in operation.

The pumps were particularly reliable, operating for 20,000 hr without maintenance.

Operating personnel recognize from their experience that the peak of the distribution of incidents occurs soon after any maintenance period. Once a disturbance from steady operating conditions occurs, faults become apparent.

4.6.3.3 *Difficulties after Start-Up*[†]

a. *Zero Doppler coefficient.* The Doppler coefficient reduced to zero after 150 days at power due to fuel expansion within the central fuel void. It is expected that it would be available under high power conditions and would still act as a safety factor. However this supposition cannot be tested.

b. *Sodium expulsion* (*39*). Figure 4.32 illustrates the scheme by which argon was moved from tank A to tank B. During the movement valve C was inadvertently left open and argon was compressed into the reactor vessel. The vessel safety valves were not designed for the high pressure that ensued, and they had a much lower limit setting, but their design caused them to fail to open at the higher value. The pressure then forced sodium out through a dip guide tube onto the reactor face. The pressure reached 0.6 bar.

About one cubic meter of sodium was ejected and fortunately it did not

---

[†] See Denielou (*37*).

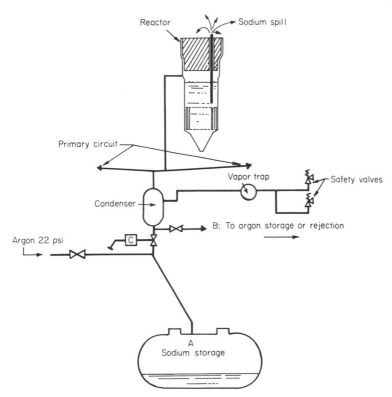

Fig. 4.32. Schematic diagram of the path of the sodium in the Rapsodie incident (*37, 39*).

ignite but simply glowed a rosy red, being hot. There was no smoke. A considerable clean-up job resulted.

Besides the obvious faults of the bad system of pumping argon from tank A to B while in connection with the vessel, and the bad design of the vessel safety valves, this incident was compounded by several other factors: There was no hole in the upright dip tube to equalize pressures inside and out. The cap had been left off because two people from different groups were responsible for the job and each left it to the other. Valve C was left open because the design of the control layout made it difficult to see the indicator which showed whether the valve was closed or not. In addition, administrative delays occurred in making a safety modification to correct the situation. These took the following sequence (*37*):

October 4, 1965. A fire was experienced at the reactor mock-up due to this same cause of sodium expulsion through a dip tube.

October 13, 1965. The mock-up operations group reported the incident.

November 22, 1965. A safety meeting analyzing the incident concluded that bad design was the cause and that holes should be drilled in the dip tube. A recommendation to Rapsodie to this effect was made.

February 4, 1966. Rapsodie group reported that, although inconvenient, a hole could be made.

March, 1966. Instructions were issued to this effect to the contractors. In the same month the contractors refused and made a counter suggestion.

April, 1966. Rapsodie operators insisted on their instructions.

During the summer discussions ensued between the operators and the contructors without resolution.

October 18, 1966. The incident occurred at Rapsodie on the day prior to a visit by the Minister of Technology. Soon after all dip tubes were cut off without further delay. However...

April, 1967. The same accident occurred in another facility.

The safety lessons inherent in this sequence of events are clear.

c. *Secondary circuit leak (39).* During filling of a secondary circuit, a filling line became plugged due to an erroneous order of heating. The heating was insufficient to avoid freezing. However, elsewhere in the circuit other insufficient trace heating caused another plug to occur independently. Between the two plugs excessive heat caused pressures to rise to give a leak. Indication of the leak into the double containment was received but ignored as it was obtained on a panel of other known faulty instruments. Thus further leaks were caused throughout the doubly contained pipework. There was no drainage point within the double containment.

d. *Fuel handling incident.* Figure 4.33 illustrates diagrammatically the operation of the hold-down tube in holding and slightly spreading assemblies around the one which is to be removed. The hold-down tube is kept in place by compressing a 0.5 ton spring to cock a latch in order to restrain a 0.3 ton upward force on the assembly to be removed.

In order to compress the 0.5 ton spring, it was convenient but bad design to use the weight of the shielding on the refueling machine that weighed 40 tons. This had the effect of overcompressing the spring. On this particular occasion, it did so and was unlucky enough to catch the edge of the hold-down tube against the marking ICZ on the top of a neighboring assembly. The horizontal edge actually caught against the upper horizontal bar of the Z. This had the effect of bending the adjacent assembly head over as shown in Fig. 4.33, so that on the removal of the hold-down tube, the hold-

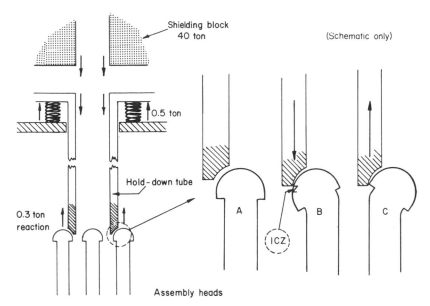

Fig. 4.33. Course of incident in which the Rapsodie hold-down tube damaged the head of an adjacent assembly on which the engraved letters were ICZ (*37*).

down tube was able to pick up the adjacent assembly by the bent head. It did exactly the thing it was designed not to do.

The procedure that used excessive weight to compress the 0.5 ton spring was bad, but the incident points toward the need to remember that every event may not be anticipated (especially those associated with a Z engraving on an assembly!).

e. *Pump intermittent operation.* Initially a pump had been jammed by extraneous material that had to be cleared, but in general there was very little difficulty with the pumps. However, in one instance during operation, flow loss occurred for no apparent reason. The reactor was shut-down, and later the pump came on to full flow in 3.5 min. Then later it again stopped, this time after a few seconds. The reason was difficult to isolate.

Figure 4.34 indicates the brushes which had been sparking. This overheated the holder which removed the brushes by differential expansion. When the pump cooled down, the brushes recontacted and the pump restarted. Visual checks did not catch this effect even during inspection. The problem was cleared by collecting the sparking on a dc collecting mechanism used as a sensing device.

f. *Plugged argon lines.* During operation, the argon lines quickly plugged

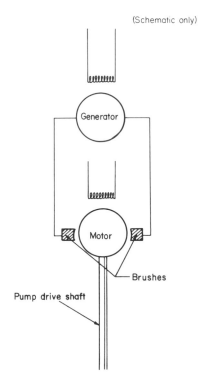

(Schematic only)

Fig. 4.34. Diagrammatic arrangement of brushes in the Rapsodie pump drive motor (*37*).

with sodium oxide crud. This was solved by pumping sodium through the argon system by an EM pump. The occurrence shows that such lines should be designed for flushing by sodium and should be used to pump sodium through from time to time.

g. *Jammed rotating plug.* The plug was exposed to sodium vapor that froze and gave rise to sufficient crud to jam the rotating plug. This problem was very difficult to solve and needed several attempts, because it got progressively worse during operation. Continuous rotation was first tried. Heating of the joint in order to keep the sodium from freezing failed. Shaping the plug fitting to avoid hold-up ledges, on which the crud might stick, did not work either.

The problem was finally solved by forcing helium down between the plug and the plug support to keep the argon and the sodium vapor contained in it, down in the vessel. This continuous purge also helped to cool the head.

Thus Rapsodie despite its successful operation has had a short catalog of unusual incidents from which lessons may be learned. Others may be expected before the useful life of the plant terminates (*37*).

These are postulated faults which are being monitored by the operators:

(a) If the flow meter associated with the siphon breaker were to need repair, its inaccessibility would make this extremely difficult.

(b) The pump cable connections are, so far, in close proximity, making a common mode failure a possibility following a cable fire. This could be remedied by separating the cabling.

(c) The serpentine concrete shielding consisted of concrete conventionally laid with organic interlayers (which contained chlorine). Despite the fact that this is in the nitrogen atmosphere, problems may arise since it is not known what the effect of irradiation on the organic interlayers might be. Has hydrochloric acid been produced? Will there be or is there already corrosion in that area close to the vessel? These questions show that the safety engineer's task does not end with successful operation of the plant.

4.6.4 ENRICO FERMI REACTOR[†]

Fermi is a 200 MWt reactor cooled by upward flowing sodium from three loops. The vessel is enclosed in a guard tank with a bottom inlet entering from a downcomer inside the guard tank. Figure 4.35 shows a cross section of the reactor vessel.

The uranium–molybdenum fuel is clad in zirconium. The fuel and cladding are integral, being coextruded. The blanket is stainless steel clad sodium-bonded fuel. The assemblies are rigidly held together in a birdcage with grids 2 in. apart. The core is mechanically restrained and compacted together with an upper plate and spider to hold down and bring the assemblies together. This restraint was due to the uncertainty about the EBR-I stability during the Fermi design period. The addition of further grids to help with the stiffening gave an extra pressure loss and caused a derating of the plant from 300 to 200 MW.

The control rods are sodium-cooled tubes of boron carbide. They are spring assisted to 2 g and slowed after insertion by a dashpot.

The assembly was equipped with a filter inside the inlet, but there was no protection against overall blockage. Inside the vessel there is a conical flow guide which doubles as a melt-down molten fuel deflector. It sits above a melt-down delaying system composed of zirconium-plated steel plates. Outside the vessel there is a graphite-lined catch pan.

[†] See McCarthy (40) and reference (41).

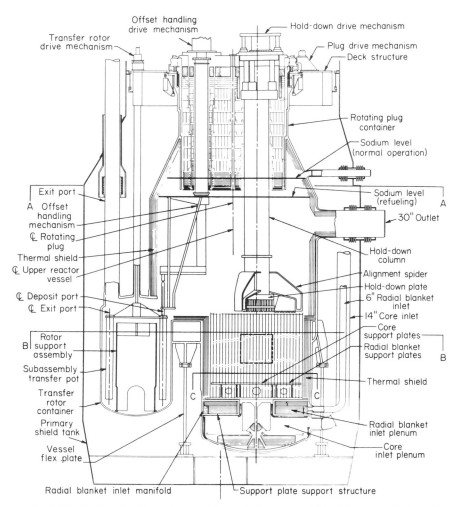

Fig. 4.35. Enrico Fermi reactor vessel, vertical cutaway *(41)*. [Courtesy of Atomic Power Development Associates, Inc.].

4.6.4.1 *Immediate Problems*

The very complex and thin vessel, with its associated fuel transfer vessel that housed a lazy Susan, caused fabrication difficulties due to its flexibility. The vessel was between one and two inches thick.

Degassing the graphite in the vessel plug gave rise to bad initial gas entrainment problems. They were solved by providing holes by which the plug could "breathe," and no gas entrainment problems have subsequently been observed.

The hold-down device is very husky, because it sees the whole of the thermal shock from shut-down to start-up and vice versa. It provides a vertical hold-down, horizontal compaction, and thermocouple support. The thermocouple information is not good, since the flow reaching the couple is obstructed by the presence of the hold-down device. Only some assemblies are sampled since there is not enough room for one thermocouple per assembly, despite the fact that one really requires at least two per assembly. However, the thermocouple operation has been very good and 80% of them still worked after two years.

4.6.4.2 *Difficulties after Start-Up*

a. *Rod incident.* Figure 4.36 shows how the control rods were operated through bellows seals used to prevent sodium vapor from reaching the rod shaft. However overpressure experiments to test the reactivity effects of pressure caused a bellows failure, and the sodium rose into the rod drive-shaft fitting. The sodium froze and caused the rod to stick immovably.

b. *IXH flow distribution.* No flow tests had been made on the heat exchanger because it was felt that the sodium, being such a good conductor of heat, would solve in itself any problems of maldistribution of flow.

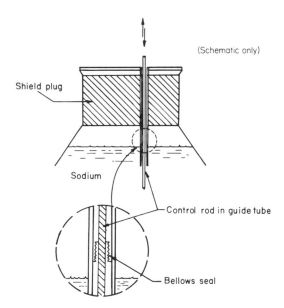

Fig. 4.36. Bellows seal failure on Enrico Fermi reactor resulting in freezing of the control rod drive shaft (*40*).

However, in fact, the flow was so badly maldistributed that the heat transfer was 40% of that expected.

$$Q = hA \, \delta T \qquad\qquad (4.53)$$

Thus as Eq. (4.53) shows, the temperature difference (δT) for a given heat flux (Q) had to be larger than design to accommodate for this failing in hA. The moral to be learned is that sodium is an ordinary working fluid and needs flow testing for any heat transfer use.

c. *Check valves.* Operation of the check valves gave sodium hammer problems which were solved by the incorporation of dashpots. These now allow up to 6% backflow, which is twice the previous value.

d. *Steam generator problems.* The steam generators had problems even before fabrication. It is a once-through design with 1200 tubes. Of the 3600 crolloy tubes delivered, one alone turned out to be carbon steel although it was stamped crolloy. This was a quality assurance problem.

During operation one tube failed and others subsequently failed due to the reaction products. The relief system worked to relieve the high pressures in the sodium side and the damaged tubes were plugged, since about 16% spare tubes are available in the design. The monitoring of the hydrogen produced by the reaction was doubtful.

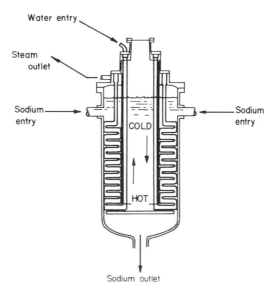

Fig. 4.37. Diagrammatic cross section of the Enrico Fermi plant steam generator design (*40*).

Figure 4.37 shows a diagrammatic cross section of one steam generator unit. Instability problems were encountered in which the top cool stagnant sodium in the central column would suddenly reverse position with the hot lower sodium. This would vary the sodium outlet temperature downward, and the automatic control reacted by cutting back the feedwater. This problem was solved by putting a lag into the control system, so that although the instability of the central column of sodium remains, the control system does not react adversely to it, but reacts only after equilibrium has been reached again and no control function is required.

e. *Refueling problems.* The shielded cask car for transferring fuel from the lazy Susan to storage and vice versa caused a number of problems. It could handle 11 assemblies and therefore it had shielding for 11 and was thus very heavy. In addition it had cooling facilities for 11 assemblies; it was self-driven; and it had argon, water, and steam lines attached. It could transfer from a core lazy Susan to a steam cleaning station, to a decay pool, or to a fresh pool port, and it was therefore far too complicated for safety. It was replaced by a simple single-assembly, single-operation machine, because it had been so troublesome. It is worth noting that the French had the same experience (Section 4.6.3).

f. *Whole subassembly blockage.*[†] A last-minute change in the design caused the addition of zirconium liners to the flow distribution cone in the reactor inlet plenum. This liner was in three sections, and two subsequently came adrift during operation when the spot-welded screws broke off. The liner sections were twisted and were forced up under the core by the coolant flow. This may have happened once toward the periphery of the core without more than a local overheating being noticed, but then, as the reactor was being brought to power, two fuel assemblies failed from overheating. This was due to the liners which had blocked the inlets to those subassemblies.

The two fuel subassemblies suffered considerable melting. See Section 4.4.4.1 for the sequence of failure that was subsequently reconstructed by analysis. It is worth now surveying the factors contributing to this accident.

4.6.4.3 *Contributing Factors in "Fermi" Melt-Down*

An indication of a small change in reactor period during start-up was ignored by the operator because it occurred at 3.00 p.m. and he had previously noticed that considerable line noise occurred at this time of the day from

---

[†] See McCarthy and Jens (27).

external power supplies, and he associated the signal on the recorder with this.

There was difficulty in determining the reactivity state of the system, or what δk had been changed during the start-up, because a staggered rod movement system was used in order to stay on a steep reactivity slope. There was no automatic computing equipment to calculate the ensuing reactivity changes.

No high temperatures were observed to indicate that sodium boiling might be occurring. This may have been because flow through those failed assemblies was very small and insufficient to transport high temperature indications from the assembly. There were not sufficient thermocouples and they were not reliable enough to give confident indications of over heating. In addition, previously "hot" assemblies had been moved and were still apparently running a little hot.

The subassemblies were eminently blockable, having simple flat orifice inlets. The inlet design has now been changed to include cruciform projections. The material to block the subassemblies was available. The last-minute design changes had not been properly documented, and diagnosis was difficult because the zirconium liners were not included on the master drawings.

There had been no acoustic "fingerprints" taken prior to the failure, and therefore acoustic detection methods could not be used for diagnosis. They could not be compared to normal steady-state and unfailed signals. There was inadequate operator monitoring analysis. However, this has now been remedied by including a computer in the diagnostic system although it is not yet integrated into the protective system.

These were the contributors to the accident which, in themselves, were all relatively innocuous but, in combination, could have a damaging potential. The value of such an incident is that the postmortem provides considerable input of data into safety engineering methods.

4.6.5 EBR-I[†]

EBR-I, shown in Fig. 4.38 in cutaway form, was a 1 MWt reactor built in 1948. It was composed of 0.5-in. diameter pins of enriched ^{235}U in a single big assembly which formed the entire 8-in. diameter core. Cooling was provided by NaK at 6 ft/sec increasing in temperature from 250 to 350°C. The vessel and piping were all doubly contained, top inlet being provided

---

† See Monson (*42*).

to the vessel, going down through the blanket, and up through the core. The coolant was pumped by EM pumps. Control was provided by bottom operated rods and reflectors for fast control with a movement of the 5 ton bottom blanket to provide slow control. This latter control suffered from poor clearances.

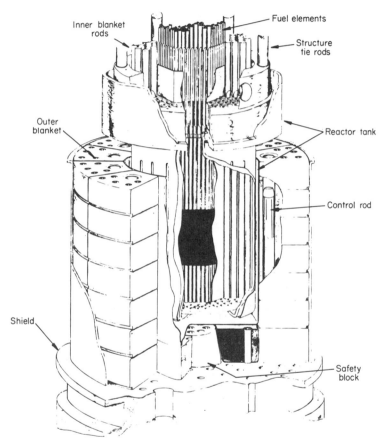

Fig. 4.38. EBR-I cutaway drawing. [Courtesy of Argonne National Laboratory.]

4.6.5.1 *Operation*

The reactor was normally stable, the power coefficient having a prompt positive component and a delayed negative component. Instabilities could occur if the power-to-flow ratio were high.

If the flow were decreased, the power would first rise and then later it would decrease to a new level. The first power changes were due to mechanical variations and bowing of fuel rods. The second changes were not understood. A new restrained core composed of rods inside a hexagon can with a tightening rod in the center and a heavy structure with which to clamp all the assemblies was provided. This new core had only a prompt negative power coefficient, having lost both the prompt positive effect of bowing and its delayed negative effect.

The eventual explanation was that, while the bowing provided a prompt inward bowing, it also brought the upper end of the fuel rod into contact with a shield plate (see Fig. 2.38) that subsequently expanded, levering the fuel rod backward away from the core and thus giving the negative effect delayed by the thermal time constant of the massive shield plate.

4.6.5.2 EBR-I *Melt-Down* (*1954*)

This melt-down occurred when doing reactivity addition runs for period tests at low flow. The order to scram was misunderstood and a slow shutdown initiated rather than the fast scram. The whole core melted down in a few seconds. The accident was of course not directly due to the above instability considerations (*43*).

The detailed sequence of events for this accident was given in Section 4.4.4.1.

4.6.6 EBR-II[†]

EBR-II, designed in 1955 under the EBR-I shadow, is the major United States irradiation facility today. It is a pool type reactor designed for 60 MWt with a metallic uranium alloy fuel but using oxide and carbide fuels in irradiation testing. It now uses oxide fuel for driver assemblies.

The core is a two-plenum design, a high-pressure inner plenum and a low-pressure outer plenum for the blankets. Emergency cooling in the 26-ft diameter tank is provided by natural circulation with an elevated IHX in the tank. The tank is doubly contained, has no fill line, penetrations, or drain lines. When natural circulation is insufficient, there is a small EM pump to augment the cooling and there are added tank coolers to reject heat by air blast coolers. These reject 10–20 kW continuously but could manage 250 kW per cooler if needed in an emergency. Also, to maintain adequate cooling, the reactor protective system does not scram the reactor flow when the reactor itself scrams. The thermal stresses inherent in this operation are accepted.

[†] See Monson (*42*), Koch (*43*), and reference (*44*).

The reactor cover is held by flexible clamps and the reactor plug has a hold-down system comprising screwed obstructions designed to take 75 psig. The reactor vault has six steel girders embedded in concrete and is designed for 300 lb of TNT. The concrete is reinforced and anchored to the containment.

The pool system was chosen here to take advantage of the simpler tank design and avoid the complex nozzles and pipes of a loop system. The pool also enjoys reliable natural circulation and a large heat capacity, although it has considerable problems associated with the head fabrication which is a vast, complex and very costly item.

Figure 4.39 shows a cutaway diagram of a vertical section of EBR-II.

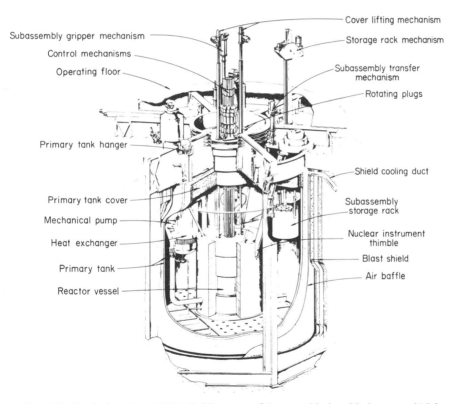

Fig. 4.39. Vertical section of EBR-II. [Courtesy of Argonne National Laboratory (*44*).]

4.6.6.1 *Operating Experience*

Thirty-three experiments or more are included in the system with a plant factor of about 45%. It has been an extremely reliable system despite the usual crop of minor incidents.

In general, the design was overenthusiastic about scram signals. There are trips on period, flux, sodium level, coolant flow rate, flow rate of change, upper plenum pressure, 12 trips of primary pumps, fire in the building, bulk sodium temperature, and many other temperatures. There are check points against changes in interlocks. As a result scrams occur two or three times a month from these 69 independent signals. Each scram wasted about 8 hr when diagnostic and return to power periods are included. Now the number of trip signals is being reduced. Excessive tripping facilities create an unhealthy and unsafe atmosphere of a nuisance protective system. Operators are under considerable pressure to keep things running and a large number of trips that are spurious create a hazardous frame of mind. On some occasions in some plants this has lead to a manual locking out of some sensors. It is far better to include power setbacks in the system and to avoid scrams on selected signals. (It is worthwhile also to note the French experience (*37*), in which 161 design primary pump trips were eventually reduced to one. This stated that the pump should be tripped if there were no flow. This trip was also removed.)

4.6.6.2 *Difficulties after Start-Up and during Maintenance*

a. *Sodium fire.* During maintenance a pipe had been plugged at two places and cut. The plugs were being kept frozen by the use of two fans, but a welder switched off one fan that was obstructing his welding operation. The plug then melted and released sodium from the open end of the pipe.

A contributor to the incident was the fact that the rest of the system was pressurized and therefore the sodium emerged under pressure. A fire ensued and a considerable clean-up job resulted.

b. *Fuel handling drop.* One step in the fuel handling procedure from the core to an in-vessel storage location was designed to give the operator a better feel for what was going on. Thus the only nonautomated part of the handling process was one designed to give the operator some psychological control.

The hold-down device spreads six adjacent assemblies, while a grapple engages the top knob of the subassembly. The operator then, remotely, engages a lifting arm (see Fig. 4.40) below the knob and closes a pin to lock-in the knob. He then performs a "wiggle" test to ensure the connection before the assembly is lifted out and transferred. In this particular case the operator did not fully engage the knob and the pin was closed with the knob out. When lifted, the knob rested on the two ends of the grapple and was lifted out of the core, over a well outside the core but inside the vessel and

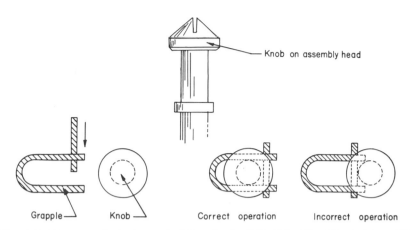

Fig. 4.40. Diagram of a fuel grapple failure in the EBR-II fuel handling incident (*43*).

then it dropped off. No damage occurred, and the assembly was easily retrieved from a previously installed catch basket in this position.

In this case, total automation might have been safer. The operator is not the most reliable of machines, and the probability of failure is unity in most cases.

c. *Lost oscillator bearings.* Bearings from an oscillator installed to do some response tests were lost from the oscillator and they jammed the control rod helical drives. However, the safety rods were not affected.

There were a number of contributory factors to this incident also. Bad design specified using ball bearings in the temporary piece of equipment (the oscillator) despite the fact that they were purposely not used in any permanent equipment. There was also a loss of quality control. The bearings were not stainless steel even though they were marked as such.

4.6.7 BR-5

BR-5 is a Soviet 5 MWt reactor built in 1959 as an irradiation facility. It is fueled with PuO_2 sintered in 0.2-in. diameter stainless steel tubes, 19 to an assembly. There are 88 such assemblies, with the fuel tubes held at the bottom but free to move at the top. The core is sodium cooled with upward flow (see Fig. 4.41) (*45*).

The system has a prompt positive contribution to the power coefficient but also has a delayed and overwhelming negative contribution.

At one time about $ 2 worth of positive reactivity was present in the core in the form of cover gas bubbles. There was some concern over safety at

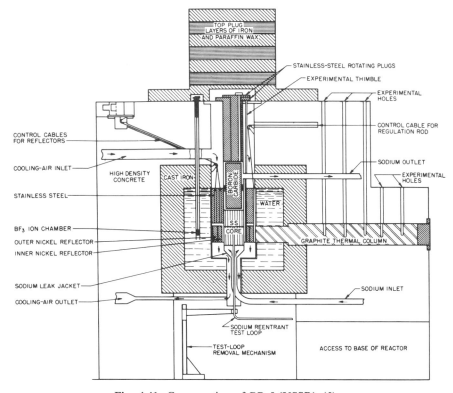

Fig. 4.41. Cross section of BR-5 (USSR) (*2*).

the time, but the problem of entrainment was cleared. There have been no other incidents. Since then some fuel cladding failures have occurred, although these are due to excessive burn-up rather than any accident event (*46a*). Despite a number of fuel cladding failures there has never been any indication of failure propagation.

4.6.8 BOR-60

BOR-60 is a second stage in the Soviet fast-reactor program, a sodium-cooled fast reactor designed to supersede the BR-5 by producing 60 MWt. It attained 20 MWt in the spring of 1970 and should reach full power early in 1971. It will provide further information for the Soviet program in which the BN-350 and BN-600 plants, 350 MWe and 600 MWe respectively, are already being built (*46b*).

The sodium for the BOR-60 fast breeder fuel irradiation reactor was shipped under a layer of paraffin wax. Some of this wax found its way into the sodium storage tanks and has thus resulted in an early clean-up problem. The experience echoed earlier purity assurance problems in Dounreay.

The plant has a conventional building which is not a leaktight containment. The design emphasizes engineered safety features, with a doubly contained primary system up to the isolation valves. Redundant safety circuits are included and backup power is provided as well as a plant designed for natural circulation to ensure that heat removal capability is available under all circumstances.

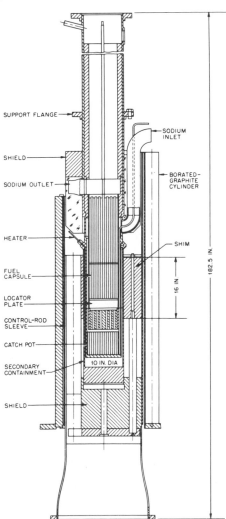

Fig. 4.42. Cross section of LAMPRE (2).

Because there is a fairly small inventory of sodium in the vessel, the outlet components experience a substantial thermal shock on scram. This is offset by a programmed flow reduction following the scram signal.

4.6.9 LAMPRE

LAMPRE was a 1 MW fast reactor built at Los Alamos in 1960. Figure 4.42 shows a cross section of the core, which was fueled by capsules of plutonium-iron alloy of about 90% plutonium. It was a cylinder about 6-in. high and 6 in. in diameter, designed to operate with molten fuel. The cooling was by upward flowing sodium (*47*).

The reactor was characterized by a reactivity loss during burn-up due to the accumulation of fission products in the fuel. About 5% volume fraction of bubbles were produced by the end-of-life. No incidents were reported. It was very stable due to the strong prompt negative fuel expansion coefficient.

4.6.10 CLEMENTINE

CLEMENTINE, a very small reactor with a power of up to 25 kW, is noted as the one reactor which was cooled by mercury. It was fueled by plutonium clad in steel with a slug of natural uranium at each end. Control of the system was achieved by moving uranium reflector and boron-10 safety rods. Figure 4.43 shows a cross section of the reactor (*48*).

Irradiation defects eventually caused the uranium plugs to swell and burst the cladding, allowing the plutonium and mercury to mix. This mixture is

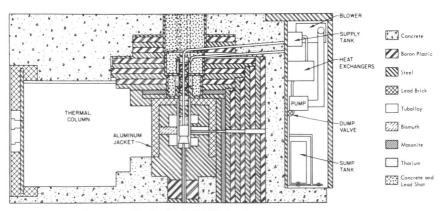

Fig. 4.43. Vertical cross section of CLEMENTINE (*2*).

a pyrophoric solution that vastly complicated the eventual disassembly of the reactor system. In addition, the uranium safety block swelled and stuck immovably.

One interesting safety highlight during the operation of the reactor concerned the possibility of damaging the system by propelling some object down the neutron window (seen to the right of the core in Fig. 4.43). Just in case, as a safety measure, the guards in the building were disarmed rather than stand the chance of one of them firing his revolver down into the core.

Safety related incidents are likely to continue; it is up to the safety engineer to learn from the errors so far committed, to collect the data from each occurrence, and to create more reliable systems.

REFERENCES

1. M. J. Driscoll, Notes on fast reactor physics. *Nucl. Power Reactor Safety Course, M.I.T., Cambridge, Massachusetts, Summer 1969.*

2. J. G. Yevick and A. Amarosi, eds., "Fast Reactor Technology: Plant Design." M.I.T. Press, Cambridge, Massachusetts, 1966.

3. J. D. Gracie and J. J. Droner, A study of sodium fires. NAA-SR-4383, Atomics International, Canoga Park, California, October 15, 1960.

4a. M. Dalle-Donne, Comparison of He, CO_2 and steam as coolants of a 1000 MWe fast reactor. *Proc. Conf. Safety, Fuels, and Core Design in Large Fast Power Reactors, Argonne Nat. Lab., October 1965,* ANL 7120. Argonne Nat. Lab., Argonne, Illinois, 1965.

4b. H. H. Hummel and D. Okrent, "Reactivity Coefficients in Large Fast Power Reactors." *Amer. Nucl. Soc.,* Hinsdale, Illinois. 1970.

4c. L. M. Olmsted, Fast breeders needed soon to conserve uranium ore. *Elec. World* **167** (24), 109 (1967).

4d. D. C. G. Smith, British fast reactors, experience and plans. Lecture. *Nucl. Power Reactor Safety Course, M.I.T., Cambridge, Massachusetts, July 1969.*

5a. M. Hori and A. Hosler, A study of gas entrainment and absorption-desorption in the Enrico-Fermi reactor and their effect on core performance. APDA-LA-6, Detroit, Michigan. July 1969.

5b. M. D. Carelli, Fuel rod design limit and transient survival criteria. WARD-4135-6, Westinghouse Advanced Reactors Division, Waltz Mill, Pennsylvania, May 1970.

6. A. M. Judd, Sodium boiling and fast reactor safety analysis. AEEW-R561, Atomic Energy Establishment, Winfrith, 1967; Boiling and condensation of sodium in relation to fast reactor safety. *Proc. Int. Conf. Safety of Fast Reactors, Aix-en-Provence, September 1967,* Paper IVa-3. Commissariat à l'Énergie Atomique, Paris, 1967.

7. R. E. Holtz and R. M. Singer, On the superheating of sodium and the generation of pressure pulses. *Proc. Int. Conf. Safety of Fast Reactors, Aix-en-Provence, September 1967,* Paper IIb-5. Commissariat à l'Énergie Atomique, Paris, 1967.

8. H. A. Bethe and J. H. Tait, An estimate of the order of magnitude of the explosion

when the core of a fast reactor collapses. UKAEA-RHM (56)/113, U. K. Atomic Energy Authority, Risley, Warrington, Lancs., England. 1956.

9. R. B. Nicholson, Methods for determining the energy release in hypothetical fast reactor meltdown accidents. *Nucl. Sci. Eng.* **18**, 207–219 (1964).

10. V. Z. Jankus, A theoretical study of destructive nuclear bursts in fast power reactors. ANL-6152. Argonne Nat. Lab., Argonne, Illinois, February 1962.

11a. D. C. Menzies, The equation of state of uranium dioxide at high temperatures and pressures. Brit. Rept. TRG Rept. 1119 (D), U. K. Atomic Energy Authority, The Reactor Group, Risley, Warrington, Lancs. England. 1966.

11b. D. Miller, A critical review of the properties of materials at the high temperatures and pressures significant for fast reactor safety. *Proc. Conf. Safety, Fuels, and Core Design in Large Fast Power Reactors, Argonne Nat. Lab., October 1965*, ANL-7120. Argonne Nat. Lab., Argonne, Illinois, 1965.

11c. W. H. Köhler, "The Effect of short delay times in Super-Prompt Critical Excursions of Fast Reactors." ANS Transactions, p. 735, San Francisco, California, Winter 1969, Nov-Dec. 1969.

12a. R. B. Nicholson, Methods for determining the energy release in hypothetical reactor meltdown accidents. APDA-150, Atomic Power Development Associates, Inc., Detroit, Michigan. December 1962.

12b. W. T. Sha and T. M. Hughes, VENUS: Two dimensional coupled neutronics-hydrodynamics computer program for fast reactor power excursions. ANL-7701, 1970.

13a. D. Okrent, AX-1: A computing program for coupled neutronics-hydrodynamics calculations on the IBM 704. ANL-5977. Argonne Nat. Lab., Argonne, Illinois, 1959.

13b. C. J. Anderson, AX-TNT: A code for the investigation of reactor excursions and blast waves from a spherical charge. USAEC Rept. TIM-951. Pratt and Whitney Aircraft, Middletown, Connecticut, September 1965.

14. J. W. Stephenson and R. B. Nicholson, WEAK EXPLOSION program. ASTRA 417-6.0, ASTRA Inc., Raleigh, North Carolina. February 28, 1961.

15. N. Hirakawa, MARS: A two dimensional excursion code. APDA-198, Atomic Power Development Associates Inc., Detroit, Michigan. June 1967.

16. T. J. Thompson and J. G. Beckerley, "The Technology of Nuclear Reactor Safety," Vol. 1, "Reactor Physics and Control," Chapter 11, p. 675. M.I.T. Press, Cambridge, Massachusetts, 1964.

17. P. J. Wattelet, Private communication, 1970.

18. A. I. Leipunskii *et al.*, Experience gained from the operation of the BR 5 reactor 1964-65. *Proc. London Conf. Fast Breeder Reactors, B. N. E. S., May 1966*, Paper 2/3, p. 171. Pergamon Press. Oxford, 1966.

19. M. D. Carelli and R. D. Coffield, Fission gas release from failed pins in liquid metal fast breeder reactors. WARD-5446, Westinghouse Advanced Reactors Division, Waltz Mill, Pennsylvania. September 1970.

20. T. C. Chawla and B. M. Hoglund, A study of coolant transients during a rapid fission gas release in a fast reactor subassembly. ANL-7651, Argonne National Laboratory, Argonne, Illinois, 1971.

21. European private Communication, 1970.

22. M. Amblard, R. Semeria, P. Vernier, A. Gouzy, and M. Najuc, Contact entre du bioxyde d'uranium fondu et un réfrigérant. Grenoble TT No. 96, Comm. à l'Energ. Atom., Grenoble, France, 1970.

23a. Argonne Nat. Lab. Reactor DevelMp. Program Progr. Repts.: ANL-7581, June 1969 ANL-7527, December 1968; Chem. Eng. Div. Res. Highlights, ANL-7550, May-December 1968; Autoclave experiments, ANL-7553, Argonne Nat. Lab., Argonne, Illinois, February, 1969.

23b. J. Graham and J. Versteeg, Consequences of Molten fuel ejection in a LMFBR, in *Proc. Am. Nucl. Soc. Nat. Topical Meeting on New Developments in Reactor Mathematics and Applications, Idaho Falls, Idaho, 1971,* to be published.

24. R. W. Tilbrook and G. Macrae, PASET: A transient code for plant analysis of sodium voiding. *Trans. Amer. Nucl. Soc.* **12,** 904 (1969).

25. H. J. Teague, Cooling failure in a sub-assembly. "An Appreciation of Fast Reactor Safety (1970)," Chapter 2. Safeguards Div., AHSB, UKAEA, 1970.

26. M. D. Carelli, H.-D. Garkisch, and J. Graham, Private communication, Westinghouse, 1970; Mark IIA subassembly failure: A Westinghouse questionnaire. UKAEA, NP-18339, Oak Ridge National Laboratory, Oak Ridge, Tennessee, 1970.

27. W. J. McCarthy and W. H. Jens, A review of the Fermi reactor fuel damage incident and a preliminary assessment of its significance to the design and operation of sodium cooled fast reactors. *Proc. Int. Conf. Safety* of *Fast Reactors, Aix-en Provence, 1967,* Paper Va-1.

28a. Private communication, 1970.

28b. J. M. Aldeanueva, N. M. Bonhomme, J. Graham, and R. G. Cockrell, Damage to a LMFBR fuel assembly duct following fuel failure. *Am. Nucl. Soc.* **14,** 287 (1971).

29. D. S. Rowe, Cross-flow mixing between parallel flow channels during boiling. Pt. I, COBRA: Computer program for coolant boiling in rod arrays. BNWL-371 (Pt. 1), Pacific Northwest Laboratory, March 1967.

30. D. S. Rowe, COBRA-II: A digital computer program for thermal-hydraulic subchannel analysis of rod bundle nuclear fuel elements. BNWL-1229, Pacific Northwest Laboratory, February 1970.

31. A. E. Walter, A. Padilla, and R. J. Shields, MELT-II: A two-dimensional neutronics and heat transfer computor program for fast reactor safety analysis. WHAN-FR-3, WADCO, High Energy Development Lab., Richland, Washington, 1970.

32a. Enrico Fermi atomic power plant: Technical information and hazards summary report, USAEC Rept. NP-11526, Pt. B, Vol. 7, DTI Extension, Oak Ridge, Tennessee. 1961.

32b. T. J. Thompson and J. G. Beckerley, "The Technology of Nuclear Reactor Safety," Vol. 1, "Reactor Physics and Control." M.I.T. Press, Cambridge, Massachusetts, 1964.

33. E. Hines, A. Gemant, and J. K. Kelley, How strong must reactor housings be to contain Na–Air reactions. *Nucleonics* **14,** 38–41 (1956).

34. J. R. Humphreys, Sodium–air reactions as they pertain to reactor safety and containment. *Proc. U.N. Int. Conf. Peaceful Uses At. Energy, 2nd, 1958,* Paper 117. U.N. Publ., New York, 1958.

35. T. S. Krolokowski, Violently sprayed sodium–air reaction in an enclosed volume. ANL-7472. Argonne Nat. Lab., Argonne, Illinois, September 1968.

36. J. Tatlock, A. G. Evans, P. K. Richards, and R. Bellinger, Design and construction of the core, reactor vessel, fuel handling equipment and shielding. *Symp. Dounreay Fast Reactor, Inst. Mech. Eng., December 7, 1960,* Paper 2.

37. G. Denielou, French fast reactors, experience and plans. Lecture. *Nucl. Power Reactor Safety Course. M.I.T., Cambridge, Massachusetts, July 1969.*

38. Rapport de Sûreté Rapsodie No. N-1007. Commissariat à l'Energie Atomique, Edition 1968.
39. G. Gajac and L. Reynes, Experience gained from the final construction phase and the approach to power of Rapsodie. *Proc. Int. Conf. Safety of Fast Reactors, Aix-en-Provence, September 1967*, Paper Va-5.
40. W. J. McCarthy, Enrico Fermi experience. Lecture. *Nucl. Power Reactor Safety Course M.I.T., Cambridge, Massachusetts, July 1969.*
41. Enrico Fermi atomic power plant technical information and hazards report, Vols. 1, 2, and 3. Power Reactor Develop. Co., Monroe, Michigan, October 1962.
42. H. O. Monson, EBR II initial operation: Highlights. *Proc. London Conf. Fast Breeder Reactors, BNES, May 1966*, Paper 2/1, p. 135. Pergamon Press, Oxford, 1966.
43. L. J. Koch, EBR I, EBR II, and 1000 MWe LMFBR studies. Lecture. *Nucl. Power Reactor Safety Course, M.I.T., Cambridge, Massachusetts, July 1969.*
44. EBR II selected illustrations distributed by ANL at the *Advisory Comm. of Reactor Safeguards Meeting, Arco, Idaho, March 17–18, 1961.*
45. V. V. Orlov *et al.*, Some problems of safe operation of the BR-5 plant. *Proc. Int. Conf. Safety of Fast Reactors, Aix-en-Provence, September 1967*, Paper Va-7. Commissariat à l'Énergie Atomique, Paris, 1967.
46a. O. D. Kazachkovskii *et al.*, Investigation of a working fuel element bundle from the BR-5 reactor with plutonium dioxide fuel. *At. Energ.* **24** (2), 136–143 (1968); translated as UDC 621.039.548.
46b. Highlights of the USSR Nuclear Power Program–1970, A condensed trip report of the USA Nuclear Power Delegation visit to the USSR, June–July 1970; WASH-1161, Divn. Reactor Development and Technology, U.S. At. Energy Comm., July 1970.
47. LAMPRE—I, Final design status report. USAEC Rept. LA-2833. Los Alamos Sci. Lab., Los Alamos, New Mexico, January 1962.
48. G. P. Arnold *et al.*, Disassembly of the Los Alamos fast reactor. USAEC Rept. LA-1575. Los Alamos Sci. Lab., Los Alamos, New Mexico, July 1953.

CHAPTER 5

CONTAINMENT

The concept of containment is to provide a series of barriers between the radioactive products of the fission process and the public. Any reactor has several such barriers, and the LMFBR has more than most. They are, successively: ceramic fuel that retains fission products; fuel-pin cladding; sodium coolant which absorbs radioactive iodine; primary circuit and vessel containment; containment building (possibly of two barrier construction); and exclusion distances.

This chapter is concerned with the latter two forms of containment: the containment building itself, which provides the final barrier to a release of radioactivity, either from normal refueling or waste removal operations or from accident conditions, and exclusion distances around the reactor itself.

5.1 Radiological Limits

Before a discussion of the functional requirements of reactor containment it is important to put radioactivity and its effects into perspective within our present environment.

5.1.1 Definition of Terms

The following are a set of definitions useful in any discussion of radioactivity:

a. *Curie (Ci)*. A measure of radioactivity. A curie of any radioactive nuclide undergoes $37 \cdot 10^9$ transformations per second.

b. *Roentgen (r)*. Named for William Roentgen, the discoverer of x rays, a roentgen is the quantity of x or γ radiation which will produce one electrostatic unit of charge in 1 cc of air at STP. It corresponds to an energy of 83 ergs.

c. *Radiation absorbed dose (rad)*. A quantity of radiation that delivers 100 ergs of energy to 1 gm of substance. In this case, body tissue is the substance of interest.

d. *Roentgen equivalent, man (rem)*. This is the biological unit: the quantity of radiation equivalent in biological damage to 1 rad of standard x rays. This unit will be used most often in this chapter.

e. *Relative Biological Effectiveness (RBE)*. This is the connection between the biological unit of the rem and the radiative unit of the rad.[†]

$$RBE = 1 \text{ rem}/1 \text{ rad} \tag{5.1}$$

f. *Linear Energy Transfer (LET)*. The average amount of energy lost per unit of particle spur-track length. The linear energy transfer of course, depends on the particle, its energy, and on the material involved. Table 5.1 shows LET values for body tissue (*1*).

The biological effect of radiation on body tissue, the RBE, is therefore related to the LET value for the particle and for the energy of that particle.

5.1.2 Damage to the Tissue

Damage to body tissue can be conveniently divided into that produced by direct and indirect action: the first being the case in which a particle of biological interest is struck by either the electron or nucleus of the ionized tissue hydrogen atom, and the other is the case in which the fragments of the ionized tissue may drift over to damage the more critical biological

---

[†] The International Commission on Radiological Protection now prefers the nomenclature Quality Factor QF for RBE when applied to protection rather than radiobiology.

TABLE 5.1

AVERAGE LET VALUES FOR PARTICLES[a]

Particle	Mass (amu)	Charge	Energy (keV)	LET[b] (keV/μ)	Tissue penetration (μ)
Electron	0.00055	−1	1	12.3	0.01
			10	2.3	1
			100	0.42	180
			1000	0.25	5000
Proton	1	+1	100	90	3
			2000	16	80
			5000	8	350
			10,000	4	1400
			200,000	0.7	300,000
Deuteron	2	+1	10,000	6	700
			200,000	1	190,000
Alpha	4	+2	100	260	1
			5000	95	35
			200,000	5	20,000

Photons are approximately equivalent to electrons of half the energy
Neutrons are approximately equivalent to protons of half the energy

[a] See Frigerio (2).
[b] Energy lost per unit track length.

molecules. Molecules which are of special interest are the DNA of the cell nucleus because damage to this molecule could have genetic consequences.

Figure 5.1 shows diagrammatically the linear energy transfer process in tissue and differentiates between the direct and indirect damage modes. Approximately, the LET values may be shown to correspond to the RBE values as shown in Table 5.2 (2).

Biological effects of body tissue damage can be separated into genetic effects, which may be transmitted to progeny, and somatic effects, which are bodily effects to the recipient itself.

5.1.2.1 Genetic Effects

It is not known whether there is a threshold to the dose required to produce genetic effects, since the data is very scant. It is difficult to obtain

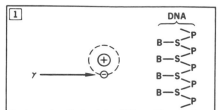

An incoming photon (γ) collides with an orbital electron (–) of one of the atoms of tissue. A DNA molecule (BSP BSP . .) rests nearby. (The letters stand for subunits of the large DNA molecule: B = a base, S = a sugar, deoxyribose, and P = a phosphate.)

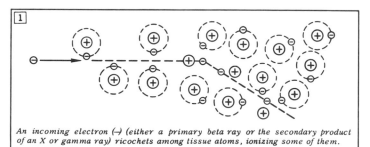

An incoming electron (–) (either a primary beta ray or the secondary product of an X or gamma ray) ricochets among tissue atoms, ionizing some of them.

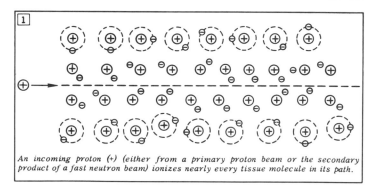

An incoming proton (+) (either from a primary proton beam or the secondary product of a fast neutron beam) ionizes nearly every tissue molecule in its path.

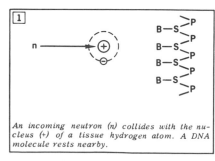

An incoming neutron (n) collides with the nucleus (+) of a tissue hydrogen atom. A DNA molecule rests nearby.

Fig. 5.1. Linear energy transfer of radiation in tissue. An illustration of direct and indirect action by high LET protons or fast neutrons and low LET electrons, or x or γ rays (2).

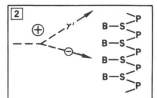

2

Energy not transferred to the electron on collision (if any), leaves as a photon of lower energy (γ'). The now energetic electron speeds off to collide with one of the sub-molecular portions of the DNA.

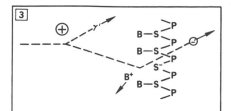

3

The bond between this portion of the DNA and the rest is split by ionization, leaving two charged fragments (B^+ and S^-) and a damaged DNA. The electron departs minus some of its energy ($-'$).

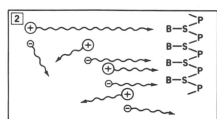

2

Many of these ionized atoms drift over to react with a nearby DNA molecule. Others drift off, eventually to combine with one another.

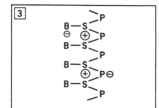

3

The ionized atoms react with the DNA, change it chemically, making it largely useless to the cell.

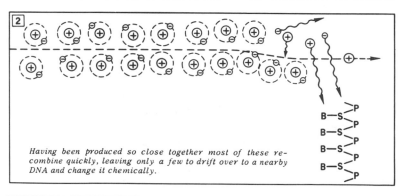

2

Having been produced so close together most of these recombine quickly, leaving only a few to drift over to a nearby DNA and change it chemically.

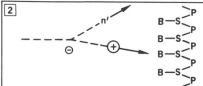

2

Having transferred some of its energy to the hydrogen nucleus (a proton), the neutron departs at a lower energy (n'). The now energetic proton collides with the DNA. Being much larger than an electron, however, it collides with many portions of the DNA, tearing it to bits.

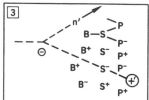

3

The ionized portions of the DNA drift away from one another, leaving the proton to continue on minus some of its energy ($+'$).

Fig. 5.1. (*continued*)

TABLE 5.2

RELATIVE BIOLOGICAL EFFECTIVENESS (RBE)[a]

Radiation type	LET[b] (keV/μ)	RBE recommended	Biological effects
x, γ, and β rays (photons and electrons) for all energies above 50 keV	0.2–3.5	1	Whole body irradiation; hematopoietic system critical
Photons and electrons 10–50 keV	3.5–7.0	2	Whole body irradiation; hematopoietic system critical
Photons and electrons below 10 keV, low energy neutrons and protons	7.0–25	5	Whole body irradiation; outer surface critical
Fast neutrons and protons, 0.5–10 MeV	25–75	10	Whole body irradiation; cataracts critical
Natural alpha particles	75–175	10	Cancer induction
Heavy nuclei, fission particles	175–1000	20	Cataract formation

[a] See Frigerio (2).

[b] Energy lost per unit track length in tissue.

results at very low dose rates, although present evidence would seem to infer that some threshold exists (3a,b).

In the meantime a linear relationship must be assumed between damage and dose. This is pessimistic and is likely to overestimate damage, although popular science writers would take the opposite view.

Radiation affects genes and gives rise to mutations, a large number of which may be undesirable (99% has been suggested). A normal person naturally experiences mutation of genes due to natural background radioactive and chemical disturbances, at a rate of about 2% to an individual. Thus for each child there is a natural 1 in 50 chance that it carries a new mutation. Extra radiation would increase this chance slowly.

Figure 5.2 shows the range of damage effects as a function of dosage. For a whole body exposure of 2 rem, the normal rate of mutation is increased by between 1.5 and 6.5% of the original value. For a whole body dose of 5 rem the normal rate is increased by between 7 and 17%, and not until a whole body exposure of over 30 rem is experienced, is the normal rate of mutation is likely to double to a total of 4% per individual (4).

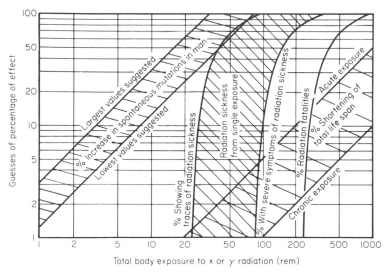

Fig. 5.2. The effect of radiation on man. Estimates of the relative effects of different doses both in the genetic and somatic ranges (*4*).

5.1.2.2 *Somatic Effects*

Whereas there is only slight evidence for a genetic damage threshold, there is good evidence for a threshold for somatic effects. It is just over 20 rem (Fig. 5.2). Below this level no somatic effects or bodily effects to the individual himself occur and above this the level of damage rises until for acute exposures, it is possible to talk of the percentage of fatalities at certain whole body doses above 100–200 rem.

A scale of values for damage in terms of doses for an average individual is the following: less than 20 rem causes no observable reactions; 30–50 rem bring detectable changes in the blood; 50–100 rem produces nausea and vomiting; 400–500 rem gives an individual a 50–50 chance of survival without medical care; and 600–1000 rem may be lethal.

5.1.2.3 *Acute Radiation Sickness*

A description of the effects of acute radiation sickness and its treatment covers the whole range of somatic effects of radiation; in less severe cases only the earlier symptoms are experienced, in more severe cases the symptoms are intensified and are usually also associated with external signs of radiation damage.

In the range of 100–500 rem the average effects may be described as follows.

Sickness:

(a) Within a dozen hours or so the patient experiences nausea, vomiting, and fatigue.

(b) After a day or two the patient begins to feel normal for a few weeks, although the blood cells (white and red corpuscles and platelets) are diminishing.

(c) The drop in red and white cells becomes obvious along with a drop in red platelets. There are two consequences: a feeling of weakness due to the anemia, and a proneness to bacterial invasion from other sicknesses as a result of the drop in the protective white corpuscles. The loss of red platelets leads to various forms of hemorrhage—from the nose, gums, or even intestines.

(d) Either the patient overcomes the blood deficiency in time or other infections, anemia, or hemorrhage lead to death.

However such sickness can be treated and the blood deficiency can be monitored and overcome in all but the most severe cases.

Treatment:

(a) It is important to first thoroughly wash and decontaminate the patient of all external and orifice radiation contamination.

(b) Then a bacterially clean environment needs to be established before stage (c) above of the sickness, to avoid infection due to external bacteria while the patient's blood count is low.

(c) It may be necessary to transfuse the constituents of blood (white and red corpuscles and platelets), and then to allow the patient to recover his normal blood-making processes after the critical period when his blood count is at a minimum.

(d) In extreme cases it may also be necessary to transfuse bone marrow which makes its way to the bone to begin its blood-forming function. The new blood-forming cells must come from a close match individual such as a twin, even though the radiation damage has also destroyed the normal rejection mechanism to strange body tissue. The rejection response will be restored as the patient recovers and it is necessary to ensure that the injected bone marrow is not subsequently rejected (5).

It is emphasized that such a sickness as is described here is the worst consequence of a radiation accident. No such accident has ever been associated with a power reactor of any type, thermal or fast. It is the intention of this book to help to ensure in some small way that such an accident will never occur in connection with a fast power reactor.

5.1.3 RADIATION LIMITS

Having described the units in which radiation is measured, and discussed how radiation relates to biological damage even in the most severe cases, it is necessary to put this information into perspective with the background radiation that is continuously received from the natural environment and then to relate to the allowable radiation doses set by regulatory bodies.

5.1.3.1 *Background Radiation*

The natural environment gives every individual a dose of approximately 100–600 mrem per year due to cosmic rays and due to radioactive materials in the Earth's crust. Table 5.3 demonstrates the origin of this natural background.

The natural cosmic ray-induced radiation varies with height above sea level, such that a resident in Denver, Colorado could better than halve his

TABLE 5.3

BACKGROUND RADIATION TO MAN[a]

Source	Dose (mrem/yr)	Possible variations
External:		
Cosmic rays[b]	30–60	Varies with altitude (at sea level: 30; at 20,000 ft: 375)
Terrestrial (^{40}K, Th, U)	30–100	Varies with location (Kerala, India: 1300; granite areas of France: 265). Varies with building construction, being highest with stone and lowest with wood
Internal:		
Th, U, and daughters	40–400	Varies with location and water supply
^{40}K in body[c]	20	Varies with the individual
^{14}C in body	2	
^{3}H in body	2	
Total:	100–600	

[a] See Wright (6).

[b] Radiation from high-altitude cosmic ray interactions.

[c] Natural radioactivity in the body.

exposure to cosmic radiation by moving to a new home at sea level. Areas at high altitude such as the Himalayan areas of Tibet are particularly high in this form of background radiation.

Terrestrial radiation also varies with location, the granite areas of France being ten times as high in radioactive background as that of the continental USA, while Kerala, India is another factor 5 higher still.

Added to this natural background are man-made radioactive sources. Table 5.4 shows the origin and levels of some man-made sources. The high rate from medical x rays is being now reduced by the education of operators, more use of shielding, and the use of discriminating x-ray beams. A great deal still remains to be done to limit the indiscriminate use of medical and dental x rays.

TABLE 5.4

MAN-MADE BACKGROUND RADIATION

Source	Annual dose rate (mrem/yr)	Assumptions
Medical x rays	50	1 medical and 4 dental/10 yr
Television	1	Black and white[a]
Watch face	25	Could be reduced by tritium paints
Weapon fallout	15	

[a] Rate is somewhat higher with color television.

In the past, highly radioactive sources were presented to the public out of an ignorance of the consequences: diagnostic x-ray machines for shoe fitting, the use of uranium compounds for coloring yellow bathtubs, further use of uranium compounds in jewelry and watch dials. Many of these sources had very high radioactive levels: the yellow pigment colored bathtubs gave an average clean individual exposure of some 100 mrem/yr. Many of these practices have now been discontinued with the imposition of modern licensing radiation standards.

Nevertheless, it is apparent that even today man may apparently be content to double his background radiation levels by his own acceptance of man-made radioactivity from such things as x rays and TV sets.

5.1.3.2 *The International Commission on Radiological Protection (ICRP)*

The ICRP has recommended limits to be placed on the dose that an individual may incur in the course of his work. These limits, published in 1967, are listed in Table 5.5 (7).

TABLE 5.5

ICRP RECOMMENDED PERMISSIBLE DOSES TO BODY ORGANS FOR OCCUPATIONAL WORKERS

Body organ	Maximum dose in any 13 weeks (rem/13 wk)	Annual permissible dose (rem/yr)	Accumulated dose to age N yr (rem)
Red bone marrow, total body, head and trunk, gonads, lenses of eyes.	3	5	$5(N - 18)$
Skin, thyroid, and bone	15	30	—
Feet, ankles, and hands	20	75	—

5.1.3.3 *Federal Regulations*

From these ICRP recommendations the AEC has set regulatory limits for the United States. These are derived from Federal Radiation Council interpretations of ICRP values.

These are set for radioactive levels due to normal nuclear plant operation and for operational transients which might be expected to occur and also for radioactive levels following accident situations. These two sets of regulatory limits are embodied in Section 10 of the Code of Federal Regulations (8), Part 20 and Part 100, respectively.[†]

a. *10 CFR 20.* This limits the exposure of individuals to radiation in restricted areas during normal operation of the plant.

(1) For the whole body, head and trunk, active blood forming organs, lenses of the eyes, and gonads: 1.25 rem/quarter.

(2) For the hands and forearms, feet and ankles: 18.75 rem/quarter.

(3) For the skin of the whole body: 7.5 rem/quarter.

A greater exposure than this may be incurred only if:

(1) During any calendar quarter the dose to the whole body does not exceed 3 rem; and

(2) the dose to the whole body when accumulated does not exceed $5(N - 18)$, where N is the age of the individual.

[†] See *Note added in proof* on p. 325.

b. *10 CFR 100*. This limits the exposure of individuals to radiation following an accident. The site will have two boundaries inside which lie an exclusion area and a low population zone, respectively. These boundaries are detailed in Section 5.2.1. The Federal regulations limit radioactivity at these boundaries in the following way.

(1) Whole body dose to an individual at the boundary of the exclusion zone during the two hours following the incident shall not exceed 25 rem.

(2) The thyroid dose at the same point shall not exceed 300 rem.

(3) The whole body dose to an individual located at the boundary of the low population zone during the whole passage of the cloud (assumed to be 30 days) shall not exceed 25 rem.

(4) The thyroid dose at the same point shall not exceed 300 rem.

(5) A distance limitation is set around the plant so that population centers of more than 25,000 persons are not involved (see Section 5.2.1).

The fast reactor contains plutonium and the 10 CFR 100 accident limits do not yet provide limits for ingested plutonium. In fact plutonium is an extremely long-lived (120-yr half-life)[†] bone seeker so that any ingested plutonium will give the individual a continuous dose throughout his lifetime. At present, limits of plutonium release are governed by ICRP limits on radiation exposures to the general public of 1.5 mrem/yr (9). Assuming a 50-year lifetime, and that ingested plutonium stays with the carrier for life, then a total of 75 mrem in any one exposure would seem an upper limit. The AEC has yet to set such a regulatory limit for plutonium.

The 10 CFR 20 limits are essentially the ICRP ones slightly undercut by taking the maximum annual permissible dose levels and dividing by four for the quarter's dose. The 10 CFR 100 doses are higher than these because of the much lower probability, and therefore frequency, of occurrence.

These limitations are, as following sections will show, met by the containment design bases with considerable safety margin. However, actual plant releases prove, in practice, to be as much as two orders of magnitude lower than even the design values for operational releases. Accidental releases have been insignificant.

5.1.4 MAIN ISOTOPIC HAZARDS

The main isotopic hazards arising from an inadvertent release from a nuclear power plant, apart from plutonium, are the iodines, kryptons, and

---

[†] This is the effective biological half-life for bone-seeking ^{239}Pu.

xenons. The iodines alone contribute to the thyroid as they alone accumulate in the thyroid gland, but all the isotopes contribute to the whole body dose.

Table 5.6 illustrates the isotopic contributions to the 30-day whole body dose for the whole passage of the cloud. The contributions to the 2-hr dose at the site boundary are of course different; ^{131}I is then the most dominant isotope instead of ^{85}Kr. These calculations assume that no iodine is absorbed by the primary sodium.

TABLE 5.6

ISOTOPIC CONTRIBUTIONS TO THE WHOLE BODY DOSE AT 30 DAYS[a]

Isotope	Percentage of total dose	Isotope	Percentage of total dose
^{85}Kr	40.1	^{135}Xe	1.8
^{131}I[b]	26.6	^{138}Xe	0.2
^{133}Xe	10.8	^{134}I	0.1
^{133}I	10.3	^{131}Xe	0.1
^{135}I	7.3	^{87}Kr	0.04
^{132}I	2.4	^{88}Kr	0.0

[a] Calculated for whole passage of cloud (30 days) at the boundary of low population zone (see Section 5.1.3.3). The calculation is dependent on release assumptions.

[b] In 2 hr dose at site boundary, ^{131}I is the dominant isotope.

In most fast reactors at this time, a particular hazard is offered by the plutonium, in the unlikely event of an accident. Plutonium is a radioactive bone-seeker with a very long half-life; thus its release is more limiting than that of any of the above isotopes. In some hypothetical accidents, it might be limiting enough to require a double containment to meet the ICRP limits.

Tritium is not a particularly troublesome hazard. The yield from fission is only 0.01%. It may be produced from the soluble boron used in the LWR as a chemical shim reactivity control, but in a fast reactor it is mainly produced in the boron carbide rods. The stainless steel cladding allows diffusion of about 80% of the tritium into the coolant stream, and this could conceivably be released to the atmosphere after passage through the primary circuit, diffusion through the IHX tubes, passage through the secondary circuit, diffusion through the steam generator tubes, and penetration to the steam and condensate circuits. If such a tortuous emission proved excessive, tritium traps could be employed in both or either of the primary and secondary circuits.

5.1.5 CALCULATIONAL METHODS

The dose at any point may be calculated according to the following equation

$$\text{dose} = (\text{total inventory}) (\text{fraction released}) \text{ BR} \int (\chi/Q) Q(t) DCF \, dt \quad (5.2)$$

where BR is the breathing rate $(3.47 \cdot 10^{-4} \text{ m}^3/\text{sec})$, DCF is the dose conversion factor (disintegrations/Ci), and χ/Q is the meteorological dispersion factor.

The release rate is $Q(t)$, taking leakage from the containment into account as well as plate-out, filtration, agglomeration, and decay removal mechanisms. Also included in $Q(t)$ is allowance for any hold-up due to a double barrier containment, so it is a sum of a set of exponentials, each with a different removal decay time constant.

$$Q(t) = \sum_{i}^{N} Q_{0i} \exp(-t/T_i) \quad (5.3)$$

The integral in Eq. (5.2) will be integrated over 2 hr for the site boundary dose and over 30 days for the low population zone boundary does. As the dose conversion factor (DCF) and $Q(t)$ are both dependent on the isotope involved, the integral will also be integrated over the isotopes of interest (the iodines for the thyroid dose and all isotopes for the whole body dose).

For a double containment design in which leakages from the inner and outer barriers are typified by the inverse time constants λ_1 and λ_2, $Q(t)$ is given by

$$Q(t) = \lambda_1\lambda_2 C_0[\exp(-\lambda_1 t) - \exp(-\lambda_2 t)]/(\lambda_2 - \lambda_1) \quad (5.4)$$

where C_0 is the inner containment fission product concentration.

The calculation of C_0 is a critical part of the dose calculation.

Codes have been produced (*10a,b*) which, starting with an initial aerosol distribution and concentration within a containment volume, model and calculate the agglomeration and plate-out of the aerosol to derive its changing concentration as a function of time (*10c*). The methods are also used in following the behavior of a sodium aerosol as a function of time. The assumption of agglomeration is critical because, for larger aerosol concentrations, the rate of agglomeration increases and the ensuing concentration at any given time thereafter reaches a maximum. Thus the dose calculated as a function of the initial aerosol concentration saturates as the agglomeration removes more and more of the initial material. Thus C_0 is obtained from these agglomeration calculations which are checked by experimental settling data.

5.1.5.1 *Results*

Following the dose calculations for particular operating and accident conditions, the containment design will be chosen to ensure that the Federal Regulations embodied in 10 CFR 20 and 10 CFR 100 are met and, indeed, are bettered.

The annual dose from a PWR from its design basis is 5 mrem/yr although it would be expected to be considerably better than that and values in the region of 0.0063 mrem/yr are likely to be experienced at the site boundary (*3b, 6*). These values should be compared to the 500 mrem/yr allowed by 10 CFR 20, and they should also be compared to the natural background of 100–600 mrem/yr and the 75–100 mrem/yr of man-made background (see Tables 5.3 and 5.4).

To put these values in perspective it is worth noting that for an increase in altitude of 600 ft there would be an increase in cosmic radiation of approximately 5 mrem/yr. Thus the federal regulations allow a release corresponding to 25,000 ft,[†] the PWR design basis allows 600 ft, but actual emissions are in the range of an equivalent 12-in. increase in altitude!

A fast reactor has no difficulty in bettering the LWR release design figures. Indeed current designs allow for essentially zero release plants during normal operation with on-site storage of fission product wastes.

5.2 Siting Limits and Considerations

The previous section has outlined the effects of radioactivity and the limits placed by the ICRP and AEC on radioactive releases from nuclear power plants. The AEC regulations are based upon maximum exposures during or following a release at points on specified site boundaries, which are also part of the regulations.

5.2.1 RADIATION SAFETY

One of the barriers mentioned at the commencement of this chapter was the physical distance which separates the public from the nuclear power plant. This barrier is defined by the following boundaries:

[†] Higher altitude changes average 300 ft/5 mrem/yr increase.

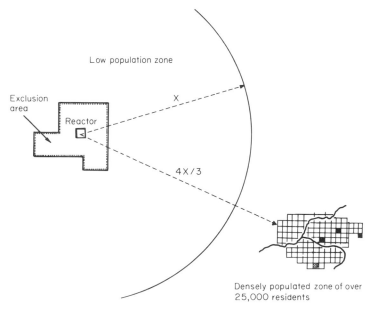

Fig. 5.3. Atomic Energy Commission site exclusion requirements for nuclear power plants.

a. *Exclusion area.* This is the area over which the reactor licensee has complete control. Residency is prohibited, or if there are residents, they are subject to immediate evacuation. There can be roads and waterways, but arrangements are made to control traffic if the need ever arises (Fig. 5.3).

b. *Low population zone.* This is the area around the plant that has a sufficiently low population so that arrangements can be made for their health and safety by evacuation or shelter. An exact population is not specified by the regulations, although the AEC would review the arrangements made for the emergency protection of residents in the zone (Fig. 5.3).

c. *Population center distance.* A population center containing more than 25,000 residents is considered a densely populated area, therefore the plant should be no nearer to such a population center than four-thirds of the low population boundary distance (Fig. 5.3).

It is clear that these boundary limits will vary for different plants in different areas. At the present time, for the LWR systems, the exclusion distance varies between 0.2 and 0.5 miles, a normal site boundary fence distance, while the low population zone varies between 1.4 and 7.4 miles. The radiation dose limits for these boundaries are set out in full in Section 5.1.3.3.

5.2.2 OTHER SITING CONSIDERATIONS[†]

Many other practical considerations determine the choice of a site of a fast reactor power plant besides the radiation limitations. The following list enumerates some siting characteristics of interest.

(a) Population density. Is sufficient unpopulated area available to meet the requirements of the previous section?

(b) Present uses of the area. Could a power plant be built in the area without disturbance to the present use of the area; for example, is it all prime agricultural or conservation land?

(c) Physical characteristics (geology, seismology, hydrology, and meteorology). Would the physical characteristics of the site make a power plant overcostly or unlicensable in view of the geological, seismological, hydrological, or meteorological difficulties?

(d) Economic location with respect to the load center. Is the site close to or far from the center of load demand or a suitable distribution grid?

(e) Proximity to water or convenience of cooling ponds or towers. Can sufficient and adequately reliable cooling water be supplied (see also Section 6.4)?

(f) Site size. Is the site large enough for any future expansion planned?

(g) Transportation facilities. Are adequate roads, railways, and waterways available for incoming plant components during construction?

(h) Labor availability. Is a labor force available in the proximity of the site to provide construction labor?

(i) Attitude of the local community. Is the local community liable to help or hinder the project? What local political situation might affect the site (see also Section 6.4)?

(j) Reactor characteristics. Does the site have any particular attractiveness for the type of reactor under consideration?

These and other similar questions all require answering before a site can be selected. Many of these questions are not safety matters, but almost all are licensing matters.

Section 5.2.4 describes a standard site, Middletown, USA, in detail and examples of answers to the above questions are given in the description. In addition, Section 6.4 discusses the licensing problems associated with present day siting in the context of disturbances to the ecology: air pollution, thermal effects, and aesthetic effects. The following section treats the effects of meteorology in more detail.

[†] See also Section 6.4.

5.2.3 Meteorology

The main meteorological concern is the wind behavior following a radioactive release or leakage following an accident. The wind dispersion appears in the dose calculations as the factor χ/Q [Eq. (5.2)]. Meteorological parameters which require study for a given site are wind direction and speed, atmospheric stability, the vertical temperature distribution, and precipitation.

5.2.3.1 *Wind Direction and Speed*

The wind direction and speed can be varied by macroscopic and local conditions. All the parameters must be measured and monitored so that not only is a mean value known, but its variances and statistical uncertainties are known also. The main characteristics are: prevailing direction that is largely a macroscopic effect dependent on the site proximity to large mountain ranges and the sea; persistence; wind shear or variation of speed and direction with height caused by the pressure gradient; local circulations due to the surface roughness in the site locality; and turbulence caused by a disparity of temperatures during the day and night.

5.2.3.2 *Atmospheric Stability*

Whether a wind condition can disperse a radioactive plume depends on whether conditions will allow the plume to be swept away, to spread out, and to eventually descend. These conditions are tied to the atmospheric stability and the temperature gradient in the atmosphere (*11*). A number of different thermal conditions contribute to this atmospheric gradient:

(a) Potential temperature decrease (lapse rate) with height is due to a temperature decrease as the pressure decreases. Assuming a dry adiabatic lapse rate without heat exchange to other air masses, there would be an upward cooling rate of 5.4°F/1000 ft. Such an adiabatic temperature condition ensures that the air will be unstable, because daytime hotter air nearer the earth is constantly changing position with upper colder air.

(b) The normal environmental lapse rate is not adiabatic, and it has some heat exchange, resulting in a lapse rate of 3.5°F/1000 ft. Superadiabatic rates give greater cooling with altitude than the normal environmental lapse rate. Also possible are subadiabatic rates giving a lesser cooling gradient. Both conditions are unstable.

(c) In some conditions the gradient can reverse itself, giving rise to inversion conditions, where the temperature increases with height. In inversion conditions the air layers are stable, cooler air remains below, and warmer air remains above. Such morning conditions can remain until the sun warms the lower regions to reverse the gradient and make the air mass unstable.

(d) In practice there may be several layers, each with a different thermal gradient.

Such wind speed and thermal stability considerations determine the wind-induced dispersal of any radioactive cloud following an inadvertent release of effluent. The wind conditions have to be characterized in terms of stability, persistence, shear, and direction. This characterization has been performed by Pasquill (*12*) and is used in dispersion calculations.

5.2.3.3 Dispersion Calculations

The dispersion depends critically on wind conditions and the stack height h from which the emission takes place. The basic dispersion equation is

$$\chi(x, y, 0) = (Q/\pi\sigma_y\sigma_z\bar{u}) \exp[-(h^2/2\sigma_z{}^2) - (y^2/2\sigma_y{}^2)] \qquad (5.5)$$

where χ is the concentration on the ground at (x, y) (Ci/m^3); Q is the release rate (Ci/sec); σ_y, σ_z are the crosswind and vertical plume standard deviations (m); $\bar{u}$ is the mean wind speed at the height of the stack h (m/sec); and x, y are the downwind and crosswind position of the point of interest relative to the stack base (m). In Eq. (5.5) the crosswind and vertical plume standard deviations are both functions of the wind conditions and are therefore tabulated as a function of the Pasquill classification. Thus from this equation the wind dispersive factor χ/Q may be calculated for insertion into the calculation of the dose by Eq. (5.2).

For a ground concentration directly downwind of the elevated source Eq. (5.5) becomes

$$\chi(x, 0, 0) = (Q/\pi\sigma_y\sigma_z\bar{u}) \exp(-h_2{}^2/2\sigma_z) \qquad (5.6)$$

Standard tables (*13*) are available for the plume standard deviations, but the actual calculation is complicated by the necessity of considering different wind conditions existing for different times. Table 6.1 details assumed wind conditions for calculations performed on the Indian Point 2 reactor plant. These conditions include allowance for building wake correction, plume meander, and various Pasquill type winds (C, D, and F) during the total time of 30 days. They give an idea of the various factors which must be included in the calculation. The set of conditions detailed in Table 6.1

are standard comparative data for PWR's rather than a set of data for the Indian Point site. They are standards set by the AEC's division of reactor licensing for use in comparative assessments.

5.2.4 MIDDLETOWN, USA

In order to standardize siting requirements for the purposes of long term 1000 MWe LMFBR reactor assessment by the industry, the AEC selected a hypothetical site (Middletown, USA) on which to place these reactors. The properties of this site give a good idea of the characteristics of any site with which a plant designer might have to contend. Many of these characteristics are safety-oriented, and others have safety overtones. The description is reprinted from the literature (*14a*). The monetary values are as of 1961 and therefore, should oney be regarded as indicating trends.

GENERAL

The Atomic Energy Commission has established Ground Rules for use in the preparation of design studies, cost normalization studies and other types of studies relating to the economic factors associated with power generation. They relate to site conditions, certain design guide data and a basis for estimating fixed charges. In the absence of specific instructions, these rules, or the applicable portions thereof, shall be used when like factual data for a specific proposal are not available. The utilization of the Ground Rules together with the AEC Classification of Construction Accounts (Section 105) will ensure to the maximum extent possible that for study and proposal evaluation purposes:

(1) The site conditions for the plant, which is designed on the basis of the Hypothetical Site, will be as uniform as possible.
(2) The estimate is prepared on a uniform Classification of Construction Accounts.
(3) The estimated indirect costs are applied to the estimated direct costs on a uniform basis.
(4) The computation of production costs is on a uniform basis.

HYPOTHETICAL SITE CONDITIONS

When location or site conditions are not specified, a selected Hypothetical Site shall be used and the plant designs and costs shall be based on the Hypothetical Site conditions described herein. The site layout and plant arrangement shall be essentially as shown in Fig. 5.4. This layout is typical for a 300 MWe nuclear power plant. Adjustments should be made as required for the proposed nuclear reactor concept.

Topography and General Characteristics

Location and Total Area. The site is located on the east bank of the North River, 35 miles north of Middletown, the nearest large city. It is 25 feet above the minimum river level and 5 feet above maximum river level. The site occupies an area of grass-covered level terrain. The land area and cost for the Hypothetical Site shall be assumed to be:

Nominal Plant Size	Land Area	Land Cost[†]
300 MWe	1,200 acres	$ 300/acre
100 MWe	1,200 acres	$ 300/acre
60–75 MWe	600 acres	$ 400/acre
50 MWe	200 acres	$ 500/acre
10 MWe	200 acres	$ 500/acre

Land is generally available surrounding the site at the same cost. It is assumed that no easements are necessary.

Access. Highway access is provided to the Hypothetical Site by a 15-mile secondary road to a state highway; this road is in good condition and needs no additional improvements. Railroad access shall be provided by constructing a railroad spur which intersects the B & M Railroad. The length of the spur from the main line to the plant site shall be assumed in accordance with the following table:

Nominal Plant Size	Length of Spur Track	Total Cost[†]
300 MWe	5 miles	$ 300,000
100 MWe	5 miles	$ 300,000
60–75 MWe	3 miles	$ 180,000
50 MWe	2 miles	$ 120,000
10 MWe	$\frac{1}{2}$ mile	$ 30,000

An airfield is located 3 miles from the state highway and 15 miles from Middletown. The North River is navigable throughout the year for boats with up to a 6-ft draft. All plant shipments will be made overland except that heavy equipment such as reactor vessel and generator stator may be barged to the site.

Population. The Hypothetical Site is near a large city (Middletown, 250,000 population) but in an area of low population density. Variation in population with distance from the site boundary is:

Miles	Population
0.25	0
0.5	60
1.0	200
5.0	2700
10.0	8000
20.0	40,000

The nearest residence to the 60–300 MWe plants is $\frac{3}{8}$ mile east of the site boundary, and to the 10 and 50 MWe plants is $\frac{1}{4}$ mile east of the site boundary, on the secondary road.

[†] Author's note: Costs are those of 1961 and, therefore, should be regarded as indicative of trends only. Absolute values today will be somewhat higher.

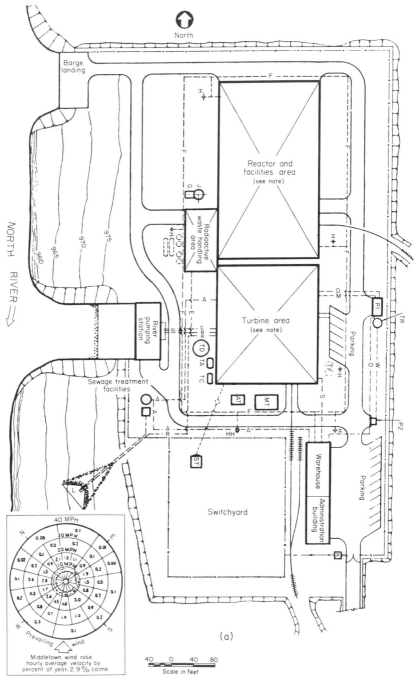

Fig. 5.4a.

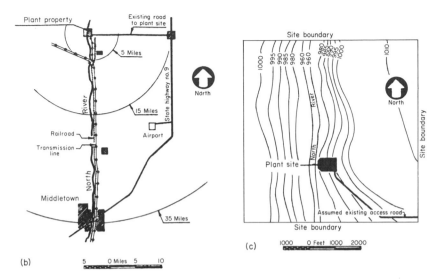

Fig. **5.4b,c.**

Fig. **5.4.** The Middletown site for the comparative evaluation of 1000 MWe LMFBRs. [Adapted from the U. S. Atomic Energy Commission (*14a*).] (a) Plot plan. The reactor and facilities area and the turbine area show the relative location of the reactor and turbine generator plants only. The detailed arrangement of reactor, turbine generator, fuel handling, and waste treatment facilities are established separately to meet the requirements of each nuclear reactor concept. Key:

D	Domestic water	TR	Raw water storage tank
M	Plant make-up water	G	Incinerator
C	Condenser water supply	J	Stack
R	Condenser water return	P2	Pump house and deep well no. 2
F	Fire protection	TF	Fuel oil tank
A	Sanitary	TD	Demineralized water storage tank
B	Service water supply	TA	Acid tank
E	Radioactive waste building effluent	TC	Caustic tank
S	Steam	MT	Main transformer
W	Well water	AT	Auxiliary transformer
H	Hydrant	K	Guard house
HH	Hose house with cart	ST	Start-up transformer
P1	Pump house, water treatment building, and deep well no. 1	L	Condenser cooling discharge seal well

(b) Location plan. (c) Site area plan.

Land Use in Surrounding Region. There are five industrial manufacturing plants within 15 miles of the Hypothetical Site. These are small plants employing less than 100 people each. Closely populated areas are found only in the centers of the small towns so the total land area used for housing is small. The remaining land, including that across the river, is used as forest or cultivated crop land, except for railroads and highways.

Utilities. Utilities are available as follows: The North River provides an adequate source of raw make-up and condenser cooling water for the ultimate station capacity. The average maximum temperature is 75°F and the average minimum is 40°F.

Natural gas service is available four miles from the site boundary on the same side of the river.

Communication lines will be furnished to the project boundaries at no cost. Cost for communication within the project boundaries will be in accordance with standard utility company practice.

Construction power is available at the southeast corner of the site boundary. Cost of this power is 15 mills per kilowatt hour.

An emergency power source in the plant is necessary, as the distribution system in the area is a single source transmission.

Meteorology and Climatology

Prevailing Wind Variation. Prevailing surface winds in the region surrounding the Hypothetical Site blow from the south through west quadrant at speeds varying from 4 to 15 miles per hour throughout the year. There are no large daily variations in wind speed or direction. Observations of wind velocities at altitude indicate a gradual increase in mean speed and a gradual shift in prevailing wind direction from southwest near the surface to westerly aloft.

Frequency of Temperature Inversions. Surface-based atmospheric inversions occur frequently during summer and early fall nights with clear skies and low wind speeds. These inversions are destroyed quickly by solar heating. Inversions occurring during winter or spring are more likely to extend into the daytime. Inversions occur most frequently when the winds blow from the south. Unstable weather conditions usually occur with winds from the north or west.

Stagnation periods with steady light winds and a high frequency of inversions are most probable from August to October. A persistent inversion with its base between 1000–4000 feet; wind speeds less than 5 miles per hour below 5000 feet; and clear skies which permit the formation of surface based inversions at night are characteristic of these periods. The annual average percentage of time with inversions is 48 percent. A survey of United States climatology records indicates that 50 percent of total annual time with inversions is a representative national average.

Frequency and Severity of Disturbances. A maximum wind velocity of 100 mph has been recorded at the site.

Snow Load. 30 psf shall be used for snow loading.

Hydrology

Precipitation. Average annual rainfall at the site is over 27 in. per year.

Drainage. Natural drainage of the site is provided by the land contours. The subterranean water travels toward the river at a velocity of 300 feet per year. The maximum temperature is 75°F with sufficient flow available to prevent exceeding the allowable temperature rise specified by the State.

Ground Water. Ground water in the region collects mostly in the weathered layer of the shale above the bedrock. Adequate ground water for sanitary supply and plant make-up is available within 50 feet below grade. Most wells in the region are drilled to the shale layer.

Geology and Seismology

Soil Profiles and Load Bearing Characteristics. Soil profiles for the site show alluvial soil and rock fill to a depth of 8 feet; Brassfield limestone to a depth of 30 feet; blue weathered shale and fossiliferous Richmond limestone to a depth of 50 feet; and bedrock over a depth of 50 feet. Allowable soil bearing is 6000 psf and rock bearing characteristics are 18,000 psf and 15,000 psf for Brassfield and Richmond strata, respectively. No underground cavities exist in the limestone.

Seismology. This is a Zone 1 site, as designated by the Uniform Building Code, based on the observation of three earthquakes of seismic intensities 6–8 on the Rossi–Forel scale in the period 1870–1958, causing minor damage to towns in the surrounding area.

Radioactive Waste Disposal

Sewage. All sewage must receive primary and secondary treatment prior to dumping into the North River.

Volatile Wastes (Radioactive and Toxic Gas). Maximum permissible concentrations or dosages shall be as prescribed in:

(a) AEC Standards for Protection Against Radiation, as published in the Federal Register (24 F. R. 8595, September 1960 and as amended in 25 F. R. 13952, December 30, 1960).
(b) National Bureau of Standards Handbook 69, Maximum Permissible Body Burdens and Maximum Permissible Concentration of Radionuclides in Air and in Water for Occupational Exposure.

In the event of conflict between items (a) and (b) above, item (a) shall govern.

Liquid Wastes. Maximum permissible activity of water entering the North River shall be as prescribed in the references listed under "Volatile Wastes" above. The activity level of the liquid effluent shall be measured as it leaves the plant. No credit for dilution in the North River will be assumed.

Solid Wastes. Storage on site for decay will be permissible but no ultimate disposal on site will be made.

Labor

Availability of Local Labor Force. Labor availability for plant construction and operation at this site is adequate, although the distance of 35 miles to the nearest large center of population requires an additional transportation allowance in the wage rates for all classes of construction labor.

Labor Productivity. Assume productivity will be equivalent to western Massachusetts.

Labor Rates. Assume craft labor rates to be in accordance with those shown by current Statistics of United States Bureau of Labor for a western Massachusetts site.

Work Week. The construction work week will be based on a 40-hour week with no regularly scheduled overtime.

Other Site Information

The Hypothetical Site is located within the general distribution area of the Central Edison System. The cost of the step-up transformer, transmission line, and all related structures and substation equipment, required to connect the power plant to the system, will not be included in the estimates of plant construction cost. Based on projections of load growth, the system will absorb the entire station output as it becomes available.

Qualified machine shops are available in Middletown so that only minimum shop facilities are necessary at the plant.

All Hypothetical Site data not provided in these ground rules shall be consistent with a western Massachusetts plant site, provided that they do not conflict with other information contained in these ground rules.

5.3 Containment Design

Another barrier mentioned at the beginning of this chapter was the physical barrier of the containment building. It is provided as the last barrier, before the exclusion distances finally separate the public from any products of fission within the fuel. The design basis of the containment building is to ensure that any leakage from the primary system following any accident is safely contained within the limits set by the AEC (Section 5.2). With this design basis certain criteria may be set for the containment building:

(a) Radiological leakage must comply with the terms of 10 CFR 20 during normal operation and with the terms 10 CFR 100 during accident conditions.

(b) This compliance must be achieved even in conditions arising from the design basis accident that is taken to be the worst accident to which the plant could ever be subjected (see Section 5.3.1).

(c) These restrictions on leakage also apply to all parts of the containment such as air locks, penetrations, isolation valves, etc.

(d) The containment building should be capable of being tested throughout life to demonstrate that it still complies with pressure and temperature conditional lcakage rates required by radiological dose limits at the site boundaries.

Section 6.1.3 shows a more detailed set of containment criteria as an illustration of how criteria may be developed for plant systems and components.

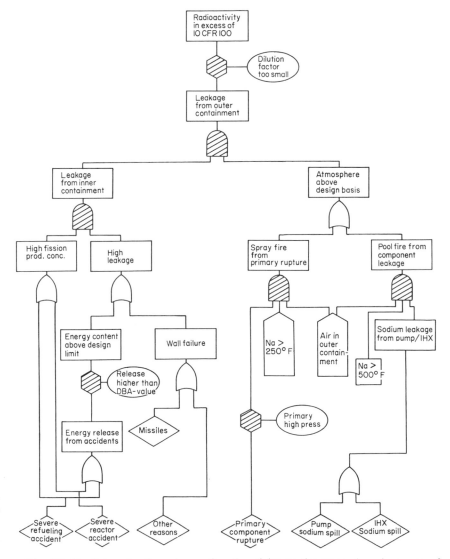

Fig. 5.5. Fault tree for the release of radioactivity to the atmosphere in excess of 10 CFR 100 limits.

5.3.1 Design Basis Accident for the Containment

Having set the limits on any radioactive release which may be permitted from the plant and having designed the plant to be accident free and so to give rise to no emissions, nevertheless it is prudent to design the containment to provide protection in the event of an accident. Thus a containment evaluation accident is chosen as the worst accident to which the plant could ever be subjected, and this accident is used to assess the adequacy of the containment design. In many cases there will be an iteration between the calculation of accident conditions and the design of the containment.

Figure 5.5 shows that for the particular containment concept chosen, a radioactive dose in excess of the limits set by 10 CFR 100 could be experienced following a core disruptive accident or a sodium fire with high radioactive contamination in the sodium only if a large number of safety features had failed or were not provided. Such a fault tree demonstrates that, for radioactive limits to be exceeded, a large number of conditions must arise at the same time. This can be seen by the large number of INHIBIT and AND gates present.

The sodium fire analysis is treated in Chapter 4 and the core disruptive accident will be treated in full in Sections 5.4 and 5.5 of this chapter. It suffices here to say that the two accidents either alone or together could give pressures in the range of 5–35 psia within the containment volume and temperatures of 200–700°F on the inner surfaces of the containment walls. Under these conditions the containment building is designed to comply with the leakage rates required to meet the dose limitations at the site boundaries.

5.3.1.1 *Other Fast Reactor Systems*

Table 5.7 shows the containment design basis accidents for fast reactors within the United States, while Table 5.8 gives details of the containment design in each case. It is notable that in each case the pressure basis for an outer steel barrier is in the range of 24–32 psig.

The thickness of a steel containment shell is calculated according to the following rules:

$$t_c = PR/(SE - 0.6P) \tag{5.7}$$

$$t_s = PR/(2SE - 0.2P) \tag{5.8}$$

where t_c and t_s are the thickness for cylindrical and spherical shells, respec-

TABLE 5.7

REACTOR DESIGN BASIS ACCIDENT CHARACTERISTICS

Reactor	Reactivity insertion ($/sec)	Total energy (MW-sec)	Work energy (MW-sec)	Accident
FARET[a]	~20	~500	60	Refueling accident Loss of coolant and failure to scram
EBR-II	200	550	440	Hypothetical: top of core falls onto bottom
Enrico Fermi	80	3300	1900	Hypothetical: top of core falls onto bottom
SEFOR	50	830	20	Hypothetical: sequential slumping of annular rings

[a] Never built.

TABLE 5.8

FAST REACTOR CONTAINMENT DESIGN

Reactor	Design bases Pressure (psig)	Temperature (°F)	Leakage[a]	Atmosphere	Construction of containment barriers
FARET	30	—	30w%/day	Depleted air	Reinforced concrete and steel liner
EBR-II	75	1200		Argon cover gas	(1) Reinforced concrete
	24	650	0.25w%/day at 20 psig	Air	(2) Steel
Enrico Fermi	32	650	0.05w%/day	Air	Steel
SEFOR	10	250	20v%/day	Depleted air and argon	(1) Reinforced concrete
	30	370	2.5v%/day	Air	(2) Steel

[a] Note that the leakage is quoted in volume percent or weight percent per day at the design basis pressure unless otherwise noted.

tively (in.); P is the design pressure (psig); S is the maximum allowable stress (psi); E is the joint efficiency; and R is the internal radius of the building (in.). Assuming a joint efficiency of unity and neglecting the small pressure effect, then, for cylindrical and spherical shells, the wall thickness (neglecting the corrosion allowance) should be $t = PR/S$ and $t = PR/2S$, respectively.

The relevant code (*13*) waives stress relief for wall thicknesses of less than one and one-eighth inches. For this thickness, the above equations show that the containment building would stand 20–30 psig. It therefore appears to be very fortunate that in each case the design basis accident gave rise to design pressures within the limit for which expensive stress relief of the containment outer shell was not required.

In practice, it is possible to include design features, such as an inner containment barrier of reinforced concrete, which would avoid subjecting the outer steel shell to high pressures even in the remote event of a core disruptive accident.

5.3.2 DESIGN

There are many possible alternative types of containment building. Some of the suggested varieties (*14b*) include:

(a) Underground containment, in which excavation alone may cost millions of dollars.

(b) Hemispherical containment with a wall and footing below grade.

(c) A prestressed concrete containment designed with compressive stresses of over 2000 psig at pouring, which can give a leakage rate below 1%.

(d) Conventional wall and roof panels for which leakage rates of less than 1% can be achieved with differential pressures of up to 0.5 psig for metal panels and 5 psig for a concrete building.

(e) Designs which include internal or external expansion volumes, such as in the case of the CP-5 at ANL which had a hold-up volume with a floating neoprene diaphragm to allow a hold-up of any fission gas release for up to 20–30 days.

As the LMFBR system contains plutonium, any gross fuel aerosol release to the containment system may necessitate leakage dilution factor of 10^{-3} or 10^{-4}. Such values cannot be achieved with a single barrier because such values cannot be tested. However it can be done with two enclosure buildings

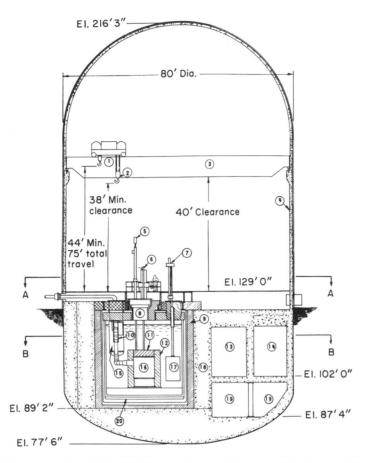

Fig. 5.6. Cross section of the EBR-II containment. [Courtesy of the Argonne National Laboratory *(14b)*.] Identification key:

1. 5-ton crane
2. 75-ton crane
3. Crane bridge
4. Concrete missile shield
5. Gripper–hold-down mech.
6. Control rod drives
7. Storage rack drive
8. Rotating plugs
9. Blast shield
10. Primary coolant auxiliary pump

11. Reactor vessel cover
12. Neutron shield
13. Basement
14. Sodium purification cell
15. Na-to-Na heat exchanger
16. Reactor
17. Subassembly storage rack
18. Concrete biological shielding
19. Subbasement
20. Primary tank

with a combined leakage of the right value, say 2 and 0.5 vol%/day or 10 and 0.1 vol%/day. Thus, for plutonium containment, two barriers are usually required although one containment volume may be merely a rather leaky aerosol settling volume.

There are many other containment design requirements. The building must house the plant with convenience and layout is of importance. The design must account for lateral stability, withstand windload, snow and roof loads, and lightning loads, and it must allow these loads to be transmitted to the foundations. It must accommodate internal mechanical and heating loads, cope with penetrations, and be designed to withstand seismic loads.

The windload, for example, is calculated according to the equation

$$F = PA_s = C_d \varrho u^2 A_s / 2g \tag{5.9}$$

where ϱ is the air density; u is the wind velocity; g is the acceleration due to gravity; A_s is the exposed surface area; P is the wind pressure; F is the wind force; and C_d is a shape factor.

The internal loads must also include those which arise from reactor transients, and particular attention must be paid to the design of air locks and various penetrations to achieve the leakage rates required.

Figure 5.6 shows the EBR-II containment building and its two barriers; the inner containment barrier comprising the reactor vessel and vaults that withstand an internal pressure of 75 psig, and the outer containment shell of steel in the form of a cylinder that would withstand at least 24 psig with an overall leakage rate of 0.25 wt%/day. Similar designs modified to include a refueling cell have been suggested for the large 1000 MWe LMFBR plants.

5.4 Design Basis Accidents

The containment as we have seen is constructed to certain design bases, one of which is, that compliance with siting radiation limits must be shown even at raised design pressure and temperature conditions within the containment. These conditions are more severe than those following the worst accident to which the plant could ever be subjected. Thus, in order to calculate the containment safety design margin, it is necessary to follow the analysis of a containment evaluation or design basis accident from initiation to its possible final consequences.

5.4.1 DESIGN BASIS ACCIDENT INITIATORS

To date, design basis or evaluation accidents for containments have been selected hypothetical occurrences. A result was postulated either in terms of gross core damage or in terms of an ultimate energy release, and in many cases an initiator was not specifically isolated as the cause of the incident.

Table 5.8 shows that EBR-II and the Enrico Fermi Reactor, both built under the shadow of the EBR-I melt-down, were evaluated on the basis of a postulated worst compaction of the core. In both cases the top of the core fell onto an already molten and slumped core, the top being presumed to have hung in position for some time before dropping.

However Table 5.8 also shows that in the case of SEFOR the first glimmerings of light were entering into containment evaluation analysis; in this case a sequential slumping of annular sections of the core was assumed.

The fact is, that although smaller experimental earlier cores could be postulated to slump in a most unrealistic manner and still give rise to only moderate energy releases, the large commercial power plants cannot assume such all-enveloping pessimism without paying a severe economic penalty in the provision of hypothetical safety features and safeguards.

Thus it is important to start with the initiator and inject some plausibility into the accident chronology and at the same time obtain a better idea of where and how safety features may be provided to prevent the occurrence of such an incident.

Therefore the present trend in containment evaluation analysis is away from the hypothetical and toward the sequential description of the accident behavior. The analyst attempts to follow the course of the core disruptive accident from initiation through to the final energy release and the distribution of that energy. In this way too he is better able to specify the required research and development to confirm his evaluation than he could have done with an upper limit analysis.

The following occurrences are incidents which might be considered to have consequences severe enough to be considered as candidates for a core disruptive accident (CDA):

(a) A total assembly blockage.

(b) A refueling accident in which an assembly is dropped into a near-critical core or in which a control rod is withdrawn or ejected from a near-critical core.

(c) A rod ejection from a critical core.

(d) Large bubbles passing through the core.

(e) A local subassembly blockage arising from a structural failure or a defective fuel pin.

(f) A pipe rupture.

(g) Scram failure in conjunction with some more probable accident such as a flow failure due to loss of power to the pumps.

5.4.1.1 *A Total Assembly Blockage*

The Enrico Fermi reactor suffered a complete blockage of two sub-assemblies as a result of a piece of structure being forced free by the coolant flow. The consequences were a melt-down of those subassemblies with some damage to surrounding subassemblies which were only partially blocked. There was no evidence that the failure had spread, although the accident occurred at low power and in an area of low power rating (*15a,b*).

A good lesson has been learned from this accident and now all LMFBR subassemblies are specifically designed against such blockage. The EBR-II nozzle design is shown in Fig. 5.7 to include side entry paths for the coolant. Most designs now include both vertical and horizontal entry paths. The only possibility for blockage could be a slow build-up of crud due to coolant contamination. Such a slow build-up could be detected either as a local flow reduction in the subassembly or by resulting higher outlet temperatures long before the blockage reached dangerous proportions. However the usual quality assurance programs also ensure that crud is of negligible quantity in the circuit. Thus a total subassembly flow blockage is not a credible CDA initiator.

5.4.1.2 *A Refueling Accident*

Experimental reactor systems operate with small shut-down margins and with, in many cases, manual refueling. Thus in college-based experimental systems a refueling fault is not unlikely. Accidents have occurred in which an assembly was dropped into a near-critical core or a control rod was withdrawn (*16*).

However, in large fast power reactors the system is shut down during refueling by as much as 10% ($ 30) and the only way in which the system could be inadvertently brought to criticality would be by removing a number of control rods completely or by misloading a number of enriched assemblies from an outer enrichment zone into che central region and then removing one or more control rods. Such a procedure would require a large number of consecutive errors on the part of refueling operating personnel.

In addition, a well designed refueling system will have interlocks and

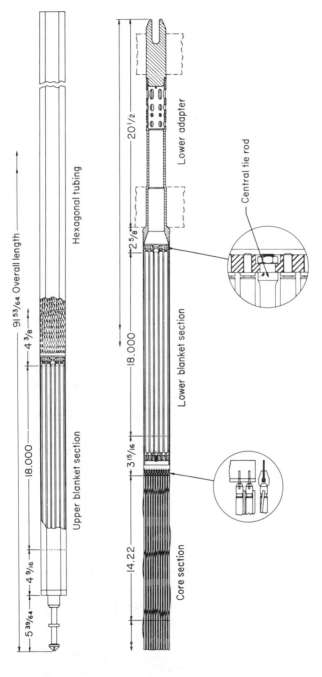

Fig. 5.7. EBR-II assembly inlet nozzle design. [Courtesy of the Argonne National Laboratory *(14b)*.]

grapple identifying features that allow the operator to positively identify what type of assembly, be it control or fuel, the refueling machine is handling at any time. Interlocks can also be included to prevent removing two control rods sequentially, and procedural controls will also prevent sequential errors.

Thus a refueling accident must be a well planned program of events and cannot be considered a credible CDA initiator in a large fast power reactor system.

5.4.1.3 *Rod Ejection*

An ejection of a control rod could conceivably give rise to very rapid rates of reactivity increase. The accident might be feasible if the control assemblies were only restrained by their own weight *and* if some ejection mechanism could be envisaged.

However, in fast reactor power systems, the control rods are generally driven into the core and their followers and drive shafts remain above the rods. Only in the case in which a rod was in the act of dropping might its upward travel be unimpeded. In other cases physical restraint would not allow more than a few thousandths of an inch of upward travel.

Suggested mechanisms for rod ejection have included coolant flashing immediately below the rod tip or simple hydraulic forces. However hydraulic forces are not sufficient to move a large control rod and coolant flashing would indicate that some other accident was already in progress and it would be most unlikely that the required force could be obtained in this way at exactly the instant of rod disengagement from its drive. Indeed, if a credible rod ejection process could be postulated, it would be relatively simple to include an anti-ejection device in the drive mechanism. However, credible ejection mechanisms cannot be envisaged, and a rod ejection cannot be considered a CDA initiator in a large fast power reactor.

5.4.1.4 *Bubbles through the Core*

A large size bubble introduced into the region of the core that has a positive void reactivity worth, could give rise to more than one dollar's worth of reactivity, and such a bubble could enter at a rate that might give rise to a very rapid reactivity rate of addition.

Section 4.2.1 discussed the sources of bubbles in some detail and showed that the likelihood of a large coherent bubble's arriving at the core inlet plenum was very remote. A bubble of tens of liters in size would be required before large reactivity ramps would result, and good design of the primary system and its cover gas surfaces prohibit the appearance of such a bubble.

Thus, although it is indeed necessary for the plant design to consider gas entrainment and accumulation very seriously, a reactivity addition from the introduction of external voids into the core is not a credible CDA initiator.

5.4.1.5 *Pipe Rupture*

A pipe rupture in a light water pressurized system is a classical accident that gives rise to immediate depressurization of the system with consequent loss of cooling and fuel failure. However, in low pressure liquid-metal-cooled systems a pipe rupture is simply a loss of cooling with an attendant loss of some coolant. The system can be adequately protected by fast shutdown of the system and by an adequate sodium inventory and emergency cooling.

Usually the accident will require that the plant be two or three cooling-loop redundant, but the accident in this low pressure system is of such low probability that it is possible to allow some localized boiling (say in the very highest rated channels) as a result. The other option of operating the reactor with additional margins to failure to cope with this particular accident may not be justified in view of the improbability of the occurrence.

Thus with a properly designed plant, the pipe rupture may lead at the most to localized boiling at the outlets of isolated subchannels and should not be considered a credible initiator for a CDA.

5.4.1.6 *Local Failures*

Local failure might include local subchannel blockages, defected fuel, or the failure of a pin that inadvertently included more highly enriched fuel than the design called for. Section 4.4 has already discussed such accidents and their possible consequences, including the possibility of a propagation of that failure to other pins and other assemblies.

The previous discussion showed that a propagation of failure from a case that resulted only in the release of some fission gases was very unlikely and even when molten fuel was involved, current analysis seemed to show that failure propagation did not occur. The most probable local defect was a subchannel blockage due to a failure of the wire wrap or grid supports or possibly from crud deposition. Such a case is the subject of much analysis and experiment aimed at determining the possible course of events and at showing that such a failure would be localized within a single assembly.

Propagation of a failure in which only gas release occurs is not considered a credible CDA initiator. A failure following a very low probability local blockage is also not expected to give rise to more than localized core damage, but this has not yet been conclusively demonstrated. No such failure propagation has ever been observed; nevertheless the accident should be retained as a low ranking candidate CDA initiator for further consideration.

5.4.1.7 *Failure to Scram*

In the higher probability accidents, such as the loss of electrical power to the pumps resulting in a reduction of core cooling, the ability to scram the reactor and to retain sufficient cooling for the removal of decay heat is necessary in order to maintain a safe system. For this reason, considerable redundancy and independence is built into the scram systems. However, if the ability to scram the reactor were lost, then a core disruptive accident would result from a loss of cooling.

The plant protective system is a multilevel redundant system: it has a multiplicity of trip signals (loss of power, loss of flow, high outlet temperature, high power-to-flow ratio), a multiplicity of circuits (redundant components and independent cables), and a redundant scram system (two independent, rapid-acting, rod shut-down systems). For this reason the failure to scram is of very low probability and has been estimated in the region of 10^{-3} to 10^{-6} per reactor year (*17, 18*). In addition, the probability of the original fault should be combined with this value for the overall probability of the total event.

Nevertheless, at this point in time, a loss of cooling accident and a failure to scram the reactor is considered a representative CDA initiator. This assumption is a direct result of licensing procedure rather than of any credibility in the accident itself, and it is of vital importance both to the fast reactor power industry and to the proper direction of safety evaluation that adequate work be done on the reliability of scram systems to show that the assumption of this accident as a CDA initiator is quite unfounded.

5.4.2 LOCAL FAILURES

The two most important local failures that may result in considerable damage to a fuel assembly are the local subchannel blockage and the pin which contains high-enrichment fuel in error and which subsequently fails in a minor transient that normal fuel could otherwise survive.

Any analysis of such incidents usually requires considerable pessimism in order to obtain a failure at all. British work has shown (*19*) that sodium boiling due to a blockage can only occur when about 8–10 channels are blocked. For blockages of more than 3 subchannels no sodium boiling occurs, although fuel pin failure may arise from overheating and consequently lead to a slow failure involving the release of fission gases which may be detected by delayed neutron monitors. If the blockage material is not inert but is composed of fissile material which can produce heat, then the situation is changed. Two adjacent blocked subchannels would produce boiling within 1 sec and the failure might propagate. The same British work showed that, if the blockages were incomplete and as little as 1% seepage of flow through the blockage existed, then boiling would not occur. In this case, extensive regions of incomplete blockage could occur without fuel damage. It would not be possible to detect the blockage from the outlet thermocouple readings unless the blockage caused severe overall flow reduction.

Thus whether fuel failure is possible as a result of subchannel blockage is critically dependent on whether the blocking material is inert or fuel, and whether the blockage is complete or not. A whole range of possible failures could result—from a small release of fission gas from a slowly developing rupture; to very much larger and quicker ruptures continuously releasing gas; to possible eventual sodium boiling. If fuel emerged, then it would probably be in the form of solid fuel that might add to the blockage, although this is most unlikely.

It should be noted that Russian experience with failed fuel (*20*) has shown that fuel does not necessarily emerge from grossly failed fuel. Similar experience was noted at the Dounreay fast reactor (*21a*).

Table 5.9 shows the possible sequence of events following a failure due to the inadvertent existence of an overenriched fuel pin at the highest rated point in the reactor (*21b*). In this case the fuel pin would have molten fuel even at normal operating power as a result of this loss of quality assurance. If the system were subjected to a minor transient which normal pins would survive within their design conditions, the fuel pin which included molten fuel might fail. The chronology here assumes that the failure is spontaneous and triggered by some unspecified disturbance. In order to analyze propagation potential, the analysis must assume not only the presence of molten fuel in the failed pin, but also some ejection mechanism. This is provided by assuming that the pin survived to high enough burn-up to produce a significant pressure due to fission product gases.

TABLE 5.9

OVERENRICHED FUEL PIN FAILURE

Time, sec[a]	Failed fuel pin behavior	Conditions adjacent
—	Hot pin is overenriched (140%) and has a proportion of molten fuel in it. A minor transient occurs and causes further fuel melting	Reactor is at full power and full flow
0.0	Molten fuel contacts cladding, which fails and releases molten fuel under fission gas pressure	—
	The ejection depends on the size of the rupture and the internal gas pressure (say 800 psi)	Molten fuel interacts with sodium and causes a pressure rise in subchannel
0.01	Molten fuel jet impingement on adjacent pin for 5–10 msec	Adjacent pin may fail due to molten fuel impingement. Results in fission gas and possibly molten fuel release directed towards the original failure
0.012	Molten fuel ejection temporarily terminates as the subchannel pressures rise due to fuel–sodium interaction	Peak voiding pressures of up to 1000 psia; subassembly duct *may* yield at corners of hexagon
0.06	—	Assembly fully voided, power decreasing due to loss of reactivity from molten fuel removal with the sodium but increasing due to sodium voiding. The reactivity change is about ±10¢. Now all pins in assembly fully blanketed by sodium vapor
0.6	—	Sodium liquid reenters and may leave again... or it may simply restore normal flow except in a few blocked subchannels

TABLE 5.9 (*continued*)

Time, sec[a]	Failed fuel pin behavior	Conditions adjacent
2.2–3.2	Further fuel pins fail due to vapor blanketing. The time of failure depends on whether the sodium reentry and chugging provides significant heat removal	—

No propagation of failure if sodium flow is restored except in a few subchannels

5.0	Fuel slumping in gross fashion under gravity as each fuel pin progressively melts and loses its cladding due to lack of heat removal only if sodium flow is not restored	Pressure pulses may increase as reentered sodium comes into contact with more sodium
10.0	Fuel fully slumped, blocking assembly. High heat fluxes to the assembly duct	Reactivity rise from slumped fuel (about 50 cents) causing the power to rise. Assembly duct *may* fail at hot spot
16.0	Heat flux to assembly duct reaches 2×10^6 Btu/hr/ft^2	Flux attains trip level at (say) 110% Reactor scrammed
17.0	Heat flux remains at full value due to stored heat	Flow reduced following scram. With high heat fluxes from failed assembly, this is the critical period during which is already subjected to pressure pulse damage might fail due to high heat input as well
20.0	Heat fluxes reducing	Flow at approximately 30%
100.0	*No propagation of failure* if the pressure pulses due to sodium vaporizing in the failed assembly did not deform the adjacent assembly critically and if adequate heat can be removed from the failed assembly by residual flow in the adjacent assembly[b]	

[a] Times are only intended to be representative of the speed at which various effects might occur. They are not all based on a particular calculation.

[b] Propagation is defined by evidence of boiling in the adjacent assembly.

Although the adjacent pin may fail due to molten fuel jet impingement, Section 4.4 has shown that this is not likely to result in propagation. A more likely total subassembly failure is that from total blanketing of the pins due to sodium vapor and fission gas release.

Nevertheless, despite high pressures due to fuel–sodium interaction that might rupture the can and deform the adjacent assembly, and despite a critical period after shut-down when the heat flux to the adjacent subassembly might be considerably in excess of the shut-down heat removal flow, the analysis shows that propagation, exhibited by boiling in the adjacent channel, does not occur. In this case this accident cannot be used to provide a CDA to be used in the evaluation of the containment.

5.4.3 FAILURE TO SCRAM

Table 5.10 illustrates the possible chronological record of a loss of cooling accident resulting from a loss of electrical supply to the pumps together with a failure to scram the reactor on receipt of signals from affected parameters.

The accident results in extensive core voiding which, in a large power reactor, results in a large positive reactivity contribution that reaches a dollar before fuel failure cladding rupture can occur. The first prompt-critical excursion is very mild and merely results in a molten fuel configuration that provides the initiating conditions for a second excursion that arises from a slumping of the fuel. The energy resulting from this second excursion is distributed throughout the vessel by shock and blast damage and damage to the head from the sodium hammer. The energy distribution of the accident will be discussed in Section 5.5.

The accident is over very rapidly (in less than 8 sec) in the representative case shown in the table. It depends entirely on the assumption that, despite the redundancy of the scram system, it nevertheless fails when required during the loss of cooling fault.

5.4.4 DESIGN BASIS ACCIDENT SUMMARY

The previous sections have discussed a number of possible CDA initiators. All of them have been shown to be most unlikely, and all but two may be ruled out as initiators. Attention would naturally be given to their absolute design prohibition. The two accidents that were not ruled out but further discussed were the local failure that might arise from an overenriched pin

TABLE 5.10

LOSS OF COOLING WITH FAILURE TO SCRAM

Time, sec[a]	Fuel behavior	Channel and power behavior
0.0	—	Normal full power. Loss of electrical supply to pumps and reduction in flow
0.9	—	Failure to scram following loss of supply signal
0.95	Fuel cladding begins to overheat	Failure to scram following low flow signal
6.5	—	Sodium boiling in hottest channels; voiding pressures about 50 psia
6.56	—	Boiling in next annulus and succeeding annuli radially outwards through core; reactivity rises
6.95	—	Prompt criticality from sodium voiding reactivity; rate $5/sec
7.3	Large regions of the fuel are molten due to the excursion. The sodium film left on the cladding vaporizes, increasing local pressures; fuel temperatures up to 3500°K	Internal pressures disassemble the core just enough to shut down. The disassembly would result in a few inches of axial movement of fuel
7.4	Fuel commences to slump under gravity depending on the distribution of molten fuel in each annulus of the core; reactivity rises	High assembly pressures due to film vaporization (about 4000 psia) maintain dry assembly conditions
7.45	—	Prompt criticality is attained at a rate of 30–50 $/sec; power excursion
7.5	Work energy release of 200–500 (say 500) MW-sec depending on the amount of sodium in the core. Pressures in the fuel about 1000 psia	Shut-down. The core fuel is dispersed in the vessel

TABLE 5.10 (*continued*)

Time, sec[a]	Fuel behavior	Channel and power behavior
7.5	Fuel and structure dispersed through the sodium above the core. Outer assemblies will be deformed by shock and will absorb some energy	Shock to vessel strains it up to 1% and relieves pressures by increased volume. Approximately 5% of energy absorbed
7.57	—	Energy transferred to sodium slug which hits the vessel plug as a hammer. Sodium velocity of order of 80–100 ft/sec
	Fuel slumps through debris, freezing and melting alternately	Hammer energy of order of 30–60 MW-sec
7.6	—	Vessel plug may lift, releasing gas

[a] Times are only intended to be representative of the speed of various effects.

or a blocked channel and the failure to scram in conjunction with another more likely event.

Analysis to date, combined with experiments on fuel–sodium interaction, transient destruction of fuel, and explosive testing of the fuel assembly wrappers, seems to indicate that propagation cannot occur following a local failure and therefore this case also should not be used to provide a CDA.

The scram failure case is analyzed as a direct result of licensing practice. It would be preferable to spend the technical effort in assessing reliability of the scram system, and this may well be done in the future. At that time a design basis for the containment may consist of a fuel handling mishap instead of a core disruptive accident. In that case the containment building will be reduced to the function of a roof for plant components.

A sodium fire is also used to evaluate containment design in many plants. Section 4.5 has shown how the sodium fire is analyzed and how its effect need not be restrictive to the design. Indeed the most sensible design solution is to inert the sodium areas so as to rule out the sodium fire entirely.

In the future, the fast reactor may well consist of a covered rather than a contained plant. This will not be possible until a breakthrough comes in the analysis of initiating events or in the experimental program carried out in support of that analysis. Such a breakthrough will have to be accompanied by a breakthrough in the licensing attitude at the same time.

5.5 Energy Partition and Mechanical Damage

The ultimate objective of any safety evaluation is to be able to specify the eventual and safe condition of the plant following any malfunction. Thus in a core disruptive accident the ultimate objective is to be able to say where the damage debris comes to rest and how this will be contained and cooled.

Thus, taking the accident described in Table 5.10 as an example, the present need is to show what damage is done by the energy release of the postulated 500 MW-sec work energy. (Section 4.3.4 gives the meaning of this work energy.)

5.5.1 SHOCK DAMAGE TO THE VESSEL

High pressures of the energy release are assumed to give rise to shock damage to the vessel in the radial and bottom head directions. The only experimental information comes from modeling and the use of scaled chemical explosive tests using TNT and pentolite. The models used included some welded, nozzled, and noncylindrical vessels to give some realism to the yield calculations (22). Some additional information may be gained from the SL-1 accident and the consequent residual strain which was measured in the vessel and core debris (23). The information is relevant, despite the fact that the SL-1 was a thermal light-water-cooled system.

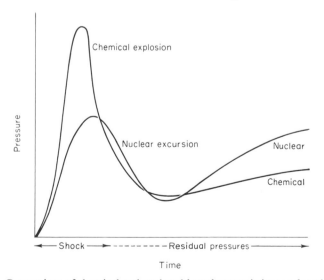

Fig. 5.8. Comparison of chemical and nuclear blast characteristics as a function of time.

Figure 5.8 shows a comparison to be expected from nuclear and chemical explosive results. The chemical explosion is faster and has a higher initial shock but a lower residual pressure than the nuclear explosion with the same integrated energy value.

Thus it is pessimistic to use the chemical explosion correlations produced for the initial radial shock deformation of the vessel. The residual pressures resulting from the explosion should be handled differently, and Section 5.5.2.2 indicates a method of reverting to a use of the total energy release for a hydrodynamic description of the excursion.

5.5.1.1 *Proctor Correlation for Shock Damage*

Work at the Naval Ordnance Laboratory produced a correlation to relate the strain in a vessel wall to the chemical explosive charge (*22, 23*).

The correlation of Eq. (5.10) expresses a relation between the shock energy W and the strain ε in terms of the dimensions of the vessel and material properties of the vessel wall

$$W = \left\{ \frac{1.407\sigma_t\varepsilon\varrho^{0.85}(3.41 + 0.117\,R_i/h_0)(R_e^2 - R_i^2)^{1.85}}{10^5(1.47 + 0.0373\,R_i/h_0)^{0.15}\,R_i^{0.15}} \right\}^{0.811} \quad (5.10)$$

where

$$\sigma_t = \sigma_y + (\varepsilon/\varepsilon_u)[\sigma_u(1 + \varepsilon_u) - \sigma_y] \quad (5.11)$$

and σ_y is the yield stress (psi), ε is the strain, ε_u is the ultimate strain, σ_u is the ultimate stress (psi), ϱ is the density (lb/ft^3), R_e is the external radius (ft), R_i is the internal radius (ft), h_0 equals $R_e - R_i$, W is the blast energy (lb TNT). Use of the correlation assumes that the vessel absorbs the strain energy radially and uniformly in the cylindrical wall, that the internal structure in the vessel has no effect, and that the classic stress–strain curve applies. The correlation applies only to radial sections of the vessel and another method must be used to assess damage to the bottom head. The top head is assumed not to see shock damage due to the large attenuation of shock pressures before arrival.

The Proctor correlation has been applied to an analysis of the SL-1 accident vessel damage to determine which observed radial damage strains could be attributed to shock and which could be attributed to blast and water hammer effects. The application provided a good representation of the observed facts and provides therefore some confirmation for the use of the correlation (*23*).

Figure 5.9 provides a plot of the SL-1 circumferential residual strain as a

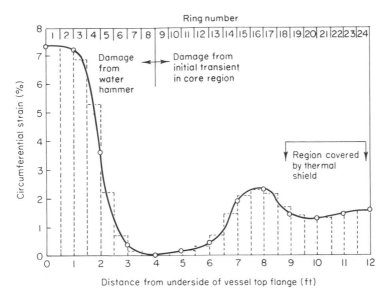

Fig. 5.9. Plot of SL-1 circumferential strain versus vessel length as measured. Data points are shown by points through which the full line curve is drawn. The dashed lines indicate the equivalent rings used in the modeling *(23)*.

function of the vessel length showing the shadowing effect of the thermal shield.

From this measured strain, the strain energy of each of the 24 rings, into which the vessel was divided, was determined by integrating the classical stress up to the strain value for each ring according to

$$SE = \tfrac{1}{6}\pi R_m th \int_0^\varepsilon \sigma \, d\varepsilon \tag{5.12}$$

where SE is the strain energy (ft-lb), R_m is the initial radius to the center of the ring (in.), t is the thickness of ring (in.), h is the height of ring (in.), σ is the classical stress in ring (psi), and ε is the strain in the ring (in./in.). This strain energy was then summed for different regions of the vessel on the assumption that different mechanisms for damage existed in each region.

It was assumed that the water coolant was cold at 300°K and that the energy release was 50 MW-sec. This energy release was equivalent to that part of the energy produced in the inner fuel elements that could be rapidly transferred to the water. The 50 MW-sec was assumed to be equivalent to 26.2 lb of pentolite producing about 459 moles of detonation gases with a pressure of 21,230 psi, calculated from the ideal gas law and a charge volume of 0.276 ft³.

Now calculating the reduction of this pressure due to the expansion of the vessel, the figure of 1780 psia was obtained and a value for the work expansion was calculated from a $P\,dv$ integration. The study assumed that the gas bubble accelerated the water above the core until it hit the vessel head, and calculated the work done in this expansion by another $P\,dv$ integration for the increase in gas volume due to the fact that it has moved the water upward. This expansion then left a residual pressure of 133 psia, and produced a hammer velocity of 147 ft/sec.

TABLE 5.11

DAMAGE ESTIMATES FOR SL-1 COMPARED WITH EXPLOSION PREDICTIONS[a]

Parameter	Explosion predictions		SL-1 damage[b]	AWRE model tests average[c]
	50 MW-sec	60 MW-sec		
Energy absorbed by lower portion of vessel[d]	2.09	2.34	1.85[e]	—
Energy in water hammer[d]	2.19	2.61	2.67	—
Total mechanical energy[d]	4.28	4.95	4.52[e]	—
Water-column velocity (ft/sec)	147	161	159	161
Impact pressure (psi)	9900	10,800	10,000	11,000
Vessel velocity (ft/sec)	26.7	29.3	29	18
Vessel rise (ft)	11.1	13.3	11.4	—

[a] See Proctor (23).
[b] See reference (25).
[c] See Warren and Rice (26).
[d] Energy in units of 10^6 ft-lb.
[e] Minimum values.

Table 5.11 shows a comparison of the damage estimates for the SL-1 and the Proctor calculations. The correlation of Eq. (5.10) was used to predict a maximum strain from the initial charge weight, on the assumption that the damage expected for a slow energy release would be similar to that from a pentolite explosion of one-eighth of the prompt nuclear energy release. The correlation predicted a strain of 0.028 in./in. against the measured maximum of 0.023 in./in.

It is recognized that using the Proctor correlation for the prediction of shock damage to the vessel wall is pessimistic in view of the different characteristics of nuclear and chemical explosions. However, Proctor (22, Table 3) does give recommended reductions in effective charge weight for slower releases.

5.5.1.2 *Stress–Strain Curve Integration*

An alternative method of assessment of vessel strain is simply to integrate the standard stress–strain curve for the work energy absorbed. This assumes uniform energy absorption, no effect from the internal structure within the vessel, and of course, that the classic stress–strain curve applies. These are the same assumptions as were made in the Proctor method. Such an integration method is:

$$W = \left\{ \frac{3\varrho(5\sigma_y + \sigma_u)\varepsilon(R_e^2 - R_i^2)^2}{3.53 \cdot 10^5} \right\}^{0.75} \tag{5.13}$$

where the units are those of Eq. (5.10). This equation also assumes, as did Proctor's, that there is a 67% efficiency in transforming blast into impulse to the vessel wall.

5.5.1.3 *Cole's Equation for Deformation Energy Absorption*

Proctor's correlation cannot be applied to deformation of the bottom head of the vessel, since it is based only on experiments that apply to the radial deformation of cylindrical vessels. Thus, for the bottom head, the following equation by Cole (*24*) may be used for damage predictions in the axial direction

$$\text{Impulse} = KW^{0.33}(W^{0.33}/R)^\beta \tag{5.14}$$

where R is the distance from the charge W, $\beta = 1.05$ for a range of underwater explosions and can be put equal to unity for the classical acoustic law with adequate accuracy, and K is an experimental factor.

Table 5.12 shows typical vessel properties that may be used in the evalua-

TABLE 5.12

VESSEL STRENGTH PROPERTIES

Property	Reactor vessel	Guard vessel
Material	SS 304	SS 304
Temperature	1000°F	500°F
σ_y	15,000 psi	18,000 psi
σ_u	45,000 psi	54,000 psi
ε_u	0.36[a]	0.40

[a] Under irradiation ε_u may decrease.

tion of these equations. The ultimate strain ε_u should be diminished by a factor of between one-third and one-half to account for vessel irregularities, welds, and nozzles. The dimunition factor is relatively constant for different riregularities considered in Proctor's explosive work (22). Thus for a real vessel

$$\varepsilon_u \text{ (for the calculation)} = \varepsilon_u/3 \qquad (5.15)$$

One large uncertainty in this type of simple calculation for damage to the vessel wall is just how much the shock is attenuated by the vessel internals. Proctor (23) has shown that the attenuation could be considerable.

5.5.1.4 *Hydrodynamic Methods*

An alternative approach is to map out the whole of the vessel internal volume as a hydrodynamic system with a two-dimensional mesh, then to represent the hydrodynamic equations of motion in each mesh node, to apply the proper boundary conditions, and to insert a pressure–temperature distribution in the core. The system can then be used to calculate the relaxation of these pressures and temperatures as a function of time in each mesh.

A series of such representations has been prepared by ANL (27a,b, 28a). These codes use the pressure–temperature distribution calculated from an energy release code such as MARS or VENUS and describe the consequences by means of a Lagrangian mesh that deforms with time.

The equations used in each mesh node are

$$\frac{\varrho}{\varrho_0} = J = \frac{dv}{dV} \qquad (5.16)$$

$$\ddot{r} = -\frac{1}{\varrho} \frac{\partial p}{\partial r}, \qquad \ddot{z} = -\frac{1}{\varrho} \frac{\partial p}{\partial z} \qquad (5.17)$$

$$dE = -p \, dV \qquad (5.18)$$

$$p = f(E, V) \qquad (5.19)$$

where ϱ, $\ddot{r}$, $\ddot{z}$, p, and E are the density, radial and vertical accelerations, the pressure, and the internal energy of the fluid, respectively. The initial volume and density are V and ϱ_0, while v is the deformed volume element.

Figure 5.10 shows a series of illustrations starting from an undeformed mesh of half the reactor system in which the left-hand boundary is considered a line of symmetry. Successive illustrations show how the mesh is deformed as a result of a sharp pressure distribution input to the mesh in the core region only. The pressure waves can be seen moving outward as a

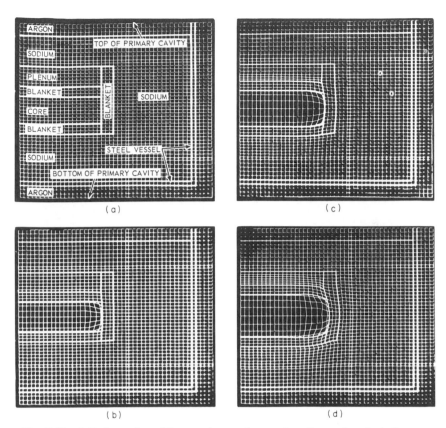

Fig. 5.10a–d. Deformation of Lagrangian mesh at various times after start of pressure pulse in a LMFBR core. Times in microseconds are: (a) $t = 0$; (b) $t = 202$; (c) $t = 322$; and (d) $t = 442$. [Courtesy of Argonne National Laboratory (*27a*)].

function of time until they contact the vessel walls first in a radial direction and then the bottom head. Figure 5.11 shows the pressure profile at the core center in an axial direction. The pressure, initially peaked, can be seen to relax into two peaks moving up and down, the one moving up being attenuated by a plenum above the core on the right, while the one moving down is held up temporarily at the vessel bottom before this finally ruptures, to relieve the pressure build-up against it. Figure 5.12 shows the radial pressure profile and the shock front moving toward the vessel wall in a radial direction.

The Lagrangian mesh deforms in the code and to retain finite difference accuracy it is necessary to rezone when the mesh gets somewhat deformed. This rezoning is a matter of experience and at present it is time consuming,

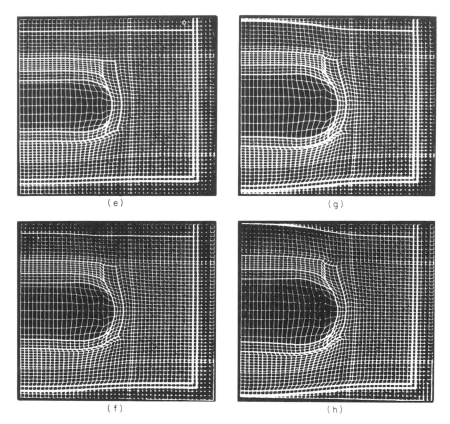

Fig. 5.10e–h. Deformation of Lagrangian mesh at various times after start of the pressure pulse in a LMFBR core. Times in microseconds are: (e) $t = 582$; (f) $t = 682$; (g) $t = 771$; and (h) $t = 810.75$. [Courtesy of Argonne National Laboratory (*27a*).]

being nonautomatic. The code has several deficiencies which are all gradually being remedied. However, it has been successfully compared to a British experiment in which a 2 oz charge of RDX/TNT 60/40 was detonated in water inside a 2-ft diameter pressure vessel. The agreement of REXCO calculated pressure values at the vessel boundary with the experimentally observed values is good (within 20%) (*28b*).

Two principal versions of the code exist, REXCO-H and REXCO-I, one being the basic code and the other including inelastic deformation of vessel walls. So far no heat transfer is included in the mesh and the model is only good for the start of an excursion following the energy release. It is therefore an excellent method of calculating radial damage, but (as is shown in the next section) for the prediction of a sodium hammer ejection above the core, heat transfer is now of importance and therefore the hydrodynamic code has to be linked to a further hammer heat transfer representation.

In principle it is to be expected that a full hydrodynamic model with heat transfer should eventually do away with the need to consider chemical explosive work.

The chemical explosive methods are, as we have seen, pessimistic. There-

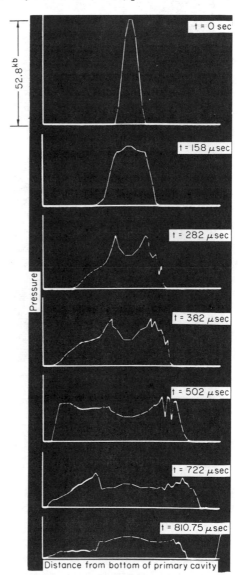

Fig. 5.11. Pressure profiles along core vertical centerline at various times after start of the pressure pulse. Times are shown in microseconds. [Courtesy of Argonne National Laboratory (*27a*).]

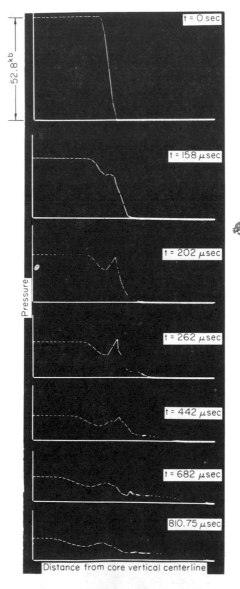

Fig. 5.12. Pressure profiles along core horizontal axis at various times after the start of the pressure pulse. Times are shown in microseconds. [Courtesy of Argonne National Laboratory (*27a*).]

fore, if they predict that a vessel will rupture, a doubt whether this will be true still lingers. However, if they predict that a vessel will not rupture then we can be confident that this is so. A 2-in. vessel with a 10-ft radius would not rupture under an explosion of 250 MW-sec if 12% strain were acceptable (Table 5.12) by this analysis. However for the same case, a more realistic evaluation of the strain would show it to be less than 1%.

5.5.2 SODIUM HAMMER EFFECTS

After the absorption of strain energy by the vessel, the remainder of the available energy is, in part, thermal energy (the core fuel is still considerably hotter than the surrounding sodium) and, in part, kinetic energy (the fuel particles are still moving outwards). The kinetic energy would be directed upward by the unruptured walls, while the thermal energy would be deposited in the sodium above the core thus vaporizing some of it. Both the kinetic energy and the sodium vaporization would combine to accelerate the sodium above the core upward toward the vessel plug.

The upward acceleration is therefore achieved by both direct momentum transfer and by further flashing of sodium as fuel particles come into contact with it. The sodium slug or hammer would eventually come to rest in contact with the vessel plug, lifting it against its restraint system. The stretching of this restraint system could allow sodium to egress from the vessel after the accident.

5.5.2.1 *Momentum Transfer*

Following the radial strain of the vessel, the chemical explosive analogy (*23*) would have us imagine that the pressure of the gases in the center of the core are relieved. Nevertheless, these pressures are still quite large and can accelerate the sodium above the core upward. The energy put into the sodium is equal to the work that is done on the hammer in moving it into contact with the vessel head.

For a 2-in. vessel with a diameter of 18 ft, an energy release of 1000 MW-sec would result in the following figures if Proctor's SL-1 analysis were followed:

Initial chemical charge equivalent to 1000 MW-sec	524	lb TNT
Charge volume	5.52	ft³
Charge volume "chemical" pressures	56562	psia
Pressure after vessel strain of 8.7%	1203.8	psia
Pressure after sodium hammer upward movement	765.0	psia
Work done on the sodium hammer	$0.204 \cdot 10^8$	ft-lb
Sodium final velocity	68.8	ft/sec

Under this sort of impact the vessel plug would suffer considerable strain. It would have a restraint system to prohibit it from becoming a missile because, if it were restrained only by its weight (say 100 tons), it could rise 92.5 ft into the air. The rise height simply serves to emphasize the potential problem. Some designs provide for bolted-on heads, rub fit heads with friction restraint, or even beam restraint.

However, this momentum calculation is based upon a chemical explosion study interested in radial deformation. The extension of this analogy to the vertical direction may be illogical, since the mechanism is much slower than the shock damage to the vessel and the postulation of an imaginary gas bubble may be quite wrong. It is therefore necessary to study the heat transfer effects by which real energy can be transferred to the sodium hammer. Proctor's SL-1 analysis (*23*) did not consider heat transfer and although the results of a momentum analysis were apparently in agreement with observed values, this may well have been fortuitous. The calculation of the upward motion of the water in the SL-1 accident analysis should be repeated with later heat transfer methods to clarify this point.

5.5.2.2 *Energy Transfer*

Recent work is directed toward deriving a model for energy transfer to the sodium by considering high energy fuel particles being ejected from the core. These are presumed to transfer energy to the sodium through which they pass, leaving a trail of vaporized sodium along their flight paths (see Fig. 5.13).

Such a model would include the three conservation equations for mass, momentum, and energy and an equation of state for the vaporization of the sodium.

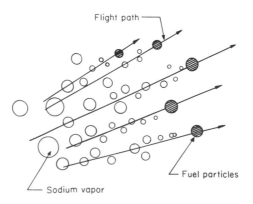

Fig. 5.13. Fuel particle flight model for transfer of heat to sodium around a disrupted core, showing sodium vapor production along the particle path.

Conservation of mass

$$\partial m_L/\partial t = -\partial m_v/\partial t \tag{5.20}$$

Conservation of momentum

$$(m_L + m_f + \tfrac{1}{2}m_v + \tfrac{1}{2}m_g)\,\partial v_L/\partial t = A(P_v - P_g) - (m_L + m_f + m_v + m_g)g \tag{5.21}$$

Conservation of energy

$$-(u_v - u_L)\,\partial m_v/\partial t = (m_v\,\partial u_v/\partial t) + (m_f\,\partial u_f/\partial t) + v_L A(P_v - P_g) \tag{5.22}$$

Equation of state

$$u_L = u_L(T) \tag{5.23}$$

In these equations m is mass, u is internal energy, P is pressure, v is velocity, A is the cross-sectional area of motion, and the subscripts L, f, v and g refer to the liquid sodium, fuel, sodium vapor, and cover gas (see Fig. 5.14).

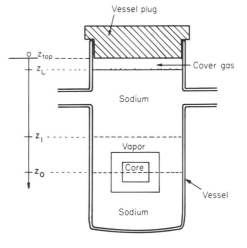

Fig. 5.14. Sodium hammer model, including the sodium vapor driving force.

The analysis requires a fuel model in order to compute the heat transferred from the fuel ($m_f\,\partial u_f/\partial t$). It assumes that incompressible sodium is being driven, by the pressure of the sodium vapor produced, against the cover gas. Such a model also needs an allowance for direct momentum transfer, since the previous driving force considered in Section 5.5.2.1 is also a driving force here. It should take account of the fact that fuel particle populations of differing sizes may be shot into the sodium in different times and the final driving force is an integrated effect of sodium vapor produced by a large number of individual energy transfer events.

Such a model could be included into a larger hydrodynamic model and could be expanded to include the effect of vessel internals. Table 5.13 shows some sample results from such a model, results which show a very moderate sodium hammer behavior, for in one case it did not even reach the vessel plug before the sodium vapor condensed. Later improved versions of this analysis included waves of particles reaching the sodium thus resulting in much larger slug energies up to 20 or 30 MW-sec (*27b*).

TABLE 5.13

SODIUM HAMMER RESULTING FROM ENERGY TRANSFER

Fuel particle radius (cm)	Particle velocity (cm/sec)	Time of arrival (sec)	Final velocity (cm/sec)
0.5	500	0.067	48.6
0.5	10000	0.119	16.7
0.05	500	No contact	0.0
0.5	500[a]	0.043	95.4

[a] The maximum pressure was held constant in the sodium vapor to simulate a wave of successive fuel particles.

5.5.3 VESSEL PLUG DAMAGE

When the slug of sodium reaches the vessel plug, an impact results and the head will be damaged in two ways: the vessel plug may deform and lift against its restraints, and lateral forces after the initial impact may cause radial deformation of the vessel just below the head.

It is important to be able to compute whether any ingress of sodium is possible into the containment area above the vessel. Therefore the plug jump and radial deformation of the vessel head need to be computed from the sodium hammer impact. Codes do exist that perform this computation: The following equations illustrate a method of deriving the plug jump from a static load analysis

$$\text{Sodium slug kinetic energy} = \tfrac{1}{2}M_s v_s^2 \qquad (5.24)$$

Momentum is conserved:

$$M_s v_s = M_T v_p \tag{5.25}$$

In these equations, M and v are mass and velocity, respectively, and the subscripts s and p refer to the sodium slug and the plug, respectively.

$$M_T = M_s + M_p \quad \text{(including shields, etc.)} \tag{5.26}$$

Assuming that the plug restraint system, say bolts, is designed to hold the plug on while absorbing the energy of the sodium slug in the stretch of the bolts themselves, we can calculate this stretch knowing the number and size of the bolts. Bolts absorb energy,

$$E_b = \text{average stress} \times \text{elongation} \times \text{volume} \tag{5.27}$$

and to absorb the total sodium slug energy

$$E_b = \tfrac{1}{2} M_T v_p{}^2 \tag{5.28}$$

From this set of equations, knowing the initial sodium slug energy, the elongation of the bolts can be derived and thus the possibility of allowing a path for the ejection of sodium may be evaluated. Note that this calculation is very pessimistic since some of the energy of Eq. (5.24) would actually be expended in the radial deformation of the vessel referred to above.

5.5.4 DESIGN EFFECTS

Damage to the head plug can be minimized to a certain extent by paying some attention to the design of the vessel and its internals.

The vessel could be allowed to rupture at any earlier stage of the accident. The SL-1 explosion interpretive tests showed that a vessel jump of 11 ft resulted from a liquid hammer if the vessel did not rupture, whereas no significant jump resulted from a case when the vessel did rupture. False rupture could be arranged in the vessel above the cover gas–liquid sodium interface by the use of carefully selected rupture disks if it seemed preferable to seek this design solution to the relief of invessel pressures.

The core barrel and thermal shields could be minimized in order to avoid the gun barrel effect whereby the force of the explosion would be directed upward. Radial vessel deformation reduces damage to the plug.

A sodium slug suppressor plate above the core could allow the plug to feel the effects of the slug movement before the sodium has time to accelerate if the suppressor plate were rigidly connected to the plug. This would allow the plug to absorb the energy gradually rather than be subjected to an impact. However the plug would also be subjected to shock, against which it would otherwise be insulated by the height of sodium and the cover gas. A European design of LMFBR does include such a sodium slug suppressor. It has not been yet demonstrated that its inclusion is an advantage.

Gas volumes included in the vessel may attenuate pressure waves, but they would have associated hazards during normal operation if they were to become connected to the primary circuit, by being a source of gas which might be introduced into the core.

To minimize the result of the impact on the vessel plug, the plug may be equipped with energy absorbing honeycomb crush shields and it may be bolted down. The head design may be such as to avoid allowing the sodium access above the operating floor by a reentry design to direct sodium to drains and from there it could be directed to vaults. Finally, outside the vessel a missile dome or shields would protect the containment from penetration.

Table 5.14 shows some of the safety features used in U.S. fast reactors to hold down the head plug and contain missiles. No system uses rupture disks, and it is debatable whether any plant should go to that extent to design for a hypothetical event.

TABLE 5.14

SAFETY FEATURES USED ON EXISTING REACTORS

Reactor	Plug hold-down	Other missile shields	Terminal cooling
EBR-II	Beams and columns in cover hold blocks on the rotating plug	Concrete missile shield	Auxiliary cooling system
Fermi	Energy absorber	Insignificant	Graphite pan beneath vessel
SEFOR	Radial beams and bolts	Yes	Bottom shield plug
FARET[a]	No plug jump if pressure less than 135 psi	Yes	No

[a] Not built but had proceeded to extensive planning before the project was canceled.

5.6 Engineered Safety Features

A list of safety features was given in Section 3.4.3 conforming to the definitions made in Section 3.4.1. Here we are interested in consequence limiting systems for various circumstances.

To avoid criticality following a hypothetical core melt-down, it could be necessary to maintain subcriticality by dispersing the fuel, by retaining the molten fuel in catchers where it might be cooled, and by providing terminal cooling to ensure that the fuel would remain where it was safe and could be cooled until the decay heat decreased.

To contain the effects of blast, head hold-down systems, blast shielding, missile barriers, and double containment are obvious design aids. Aerosol settling devices, filtration systems, and hold-up volumes would also aid the reduction of radioactive effluent following an energy release that released fuel from the system.

To avoid sodium fires and other reactions, the absence of air from the vaults and the containment, and the absence of water from these areas are self-evident solutions. The retention of sodium below its ignition temperature at those times when a sodium fire is more likely to occur (during maintenance) can be just as effective. Procedural controls are, of course, extremely important in separating sodium from oxygen and keeping it separate during start-up of the plant, during operation, and during maintenance.

5.6.1 To Avoid Criticality

In principle, it is important to maintain a subcritical system during the entire period in which the system is subjected to thermal failure, so that there is no excursion and no explosion, and to continue to maintain that subcriticality during the decay period until the damaged core can be removed.

Even if there had been a core disruption, the debris would need to be maintained in a subcritical condition to avoid further, and possibly worse, excursions. This would need to be done either until debris could be removed after the decay heat had subsided so that the fuel could freeze.

Figure 5.15 illustrates the system used in the *Fermi Reactor* (29). It comprises:

(a) a zirconium clad conical flow guide designed to disperse any molten fuel from the core into a distributed and subcritical configuration;

(b) a series of zirconium clad plates in a melt-down section through which

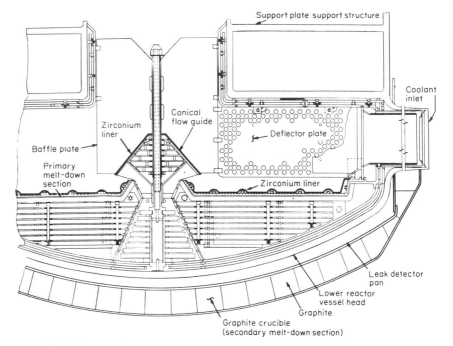

Fig. 5.15. Cross section of the inlet plenum and melt-down section in the Enrico Fermi lower reactor vessel (*29*). (Courtesy of Atomic Power Development Associates, Inc.)

the fuel could possibly melt (this melt-down system would provide a time delay to reduce the decay heat level before the fuel reached its final resting place);

(c) a further internal cone in the melt-down section to maintain sub-criticality by dispersion;

(d) the vessel and the guard vessel, both of which provide delay time while the fuel penetrates them, and provide a coolant system boundary for some sodium cooling while the fuel is melting downward; and

(e) a graphite crucible outside the vessel designed to catch and retain any molten fuel which reaches it (at this point, the molten mass would be cooled by sodium in the flooded vault).

In fact, as Section 4.6 relates, the zirconium cladding on the first cone worked loose and blocked the coolant channels, thus causing a small melt-down of two assemblies. The molten fuel from these assemblies froze in the lower section of the core.

It is worth learning the lesson that safety features in the system should also be evaluated for their possible adverse effect on safety in other respects.

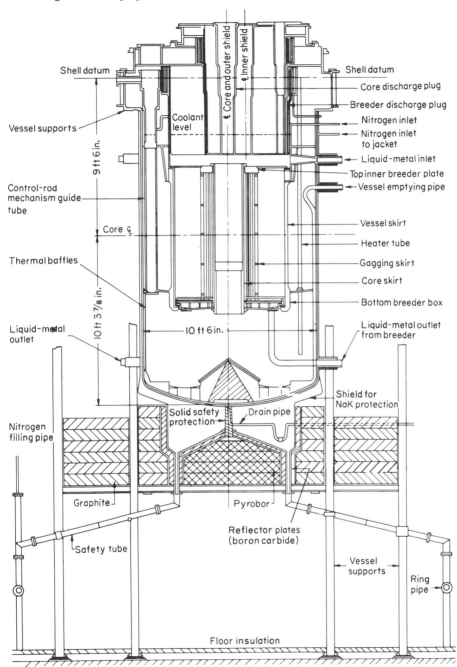

Fig. 5.16. Vertical cross section of the Dounreay fast reactor showing the core, charge machine, and the reactor vessel (*30*).

Figure 5.16 illustrates the system used in the *Dounreay Fast Reactor* (*30*).
It comprises:

(a) annular fuel elements with an inner cladding (vanadium) having a
lower melting temperature than the outer cladding (niobium) (intended to
direct any fuel release following fuel element failure down the inside of the
element itself);

(b) downflow to help molten fuel leave the core and make the system
subcritical rather than to retain it in the system (downflow forces comple-
ment gravity);

(c) an invessel cone to disperse the fuel in the same way as in Fermi;

(d) an outside vessel—a graphite, steel lined cone—again to disperse the
melt into a subcritical configuration and into its final catchpots; and

(e) twenty-four steel lined drains or catchpots to take any molten fuel
away into the bedrock on which the reactor stands.

This system has never been used, and the reactor has never experienced
more than minor fuel failures, in which no molten fuel was involved.

The modern designer faces problems very similar to these, with the added
complication that the fast reactors under consideration are now much
larger. Dounreay is 72 MWt, the Fermi plant is 200 MWt, while present
day systems in design are from 800 to 2500 MWt. With the larger plants,
even in decay mode, more heat is produced than in Fermi at full power.
This demonstrates the magnitude of the task of providing for cooling of a
damaged core even when shut down.

Scaling up the simple solutions becomes prohibitively expensive and
uncertain. It is difficult to justify such addition to the design to cope with
an accident when confronted with a growing amount of evidence saying
that such an accident will never occur.

Some recent work indicates that it may be possible to allow the molten
fuel to take its own course in penetrating the vessel and falling onto the
vault floor. Melt-through of the vessel would take about 5–10 min if the
whole core were involved, although fuel cooled by a pool of sodium above
might never penetrate the vessel if only one or two assemblies were involved.

Even if the fuel emerged from the vessel, it could be contained in from 5
to 20 ft of concrete depending on whether cooling was applied through a
cooling pipe system close to the concrete bed or not. Other factors which
are critical to such a calculation are the delay before the fuel arrives at the
concrete, what heat is being produced by decay, and, moreover, in what
form the debris arrives at the concrete. Exactly what occurs is dependent
on the type of concrete in use (*31a,b*) and it is important to have compatible

materials and concrete which does not dissociate or produce gaseous substances such as carbon dioxide which could over-pressurize the system. No design solution has yet been found or, moreover, has been shown to be needed.

Further study at ANL is showing that the heat flux from debris is such that, with some in vessel cooling, the best and most promising position in which to retain debris is above the core support plate. This has the advantage of being simpler to design to and somewhat more believeable than core catchers used to date.

5.6.2 To Contain the Effects of Blast

5.6.2.1 *Blast Shielding*

To allow the vessel to deform radially but not to such an extent that it damages the reactor cavity walls or thereby itself, blast shielding may be used effectively to strengthen the radial wall.

Figure 5.17 illustrates the radial cross section of shielding of a typical LMFBR. The shielding is provided for radiation reduction, and it may therefore be composed of serpentine concrete encased in steel. However,

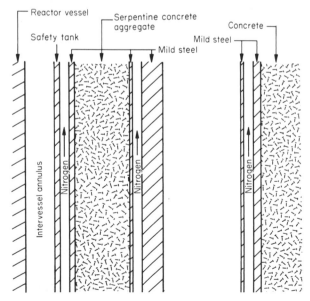

Fig. 5.17. Cross section through a typical radial neutron shield with considerable potential as blast shielding.

the total radial thickness of mild steel (7.5 in. in this case) forms a very efficient and effective blast shield, even taking no credit for other materials in the shielding system.

The analysis of the blast shield is performed using the Cole charge weight correlation (*24*):

$$I = 1.46 \ W^{0.33}(W^{0.33}/R)^{0.9} \tag{5.29}$$

where I is the specific impulse (lb-sec/in.^{2}), W is the charge weight (lb TNT), and R is the reactor shield radius (ft).

The kinetic energy is

$$KE = (IA)^2/2M \tag{5.30}$$

and the energy of strain is

$$SE = 2\pi \ \delta\sigma_a t \tag{5.31}$$

where A is the inner surface of the unit annulus (in.^{2}), M is the annulus mass (slugs), δ is the radial elongation (in.), t is the annulus thickness (in.), and σ_a is the average stress during the strain (lb/in.^{2}).

For the blast shield to perform as required the strain energy must be greater than or equal to the kinetic energy. Assuming an elongation before rupture of a given amount and summing the strain energies for each annulus of the blast shield, this sum can be compared to the calculated kinetic energy to determine whether the blast shielding is adequate.

To satisfy the model, the nitrogen-cooling annuli and the concrete annuli are assumed to have been collapsed without significant energy absorption. The shield of Fig. 5.17 can accommodate 3000 lb of TNT assuming an elongation $\delta R/R$ of 0.3. Thus it can be seen that the inclusion of radial radiation shielding will usually provide more than adequate blast shielding at the same time.

5.6.2.2 *Missile Barriers*

To protect the containment building integrity against missiles that might be generated at the reactor vessel head following a core disruptive accident, the source and strength of possible missiles must be assessed. If critical components do become effective missiles, then either barriers or adequate height may have to be provided within the containment design.

Missiles might arise from control rod drives or their shafts, from unused head penetration plugs, from head restraint bolts which snap under impulse

by the sodium slug, from whipping cables, or from small components associated with head equipment.

The energy imparted to each is calculated by a momentum transfer, and
the trajectory of the missile may be calculated by the usual mechanics. If
the trajectory meets a containment building, the velocity and energy at
impact can be calculated in order to find the penetration the missile produces.

The real characteristic of interest is the areal density in lb/ft^2, since the
missile is more penetrating if it has a large mass and small effective diameter. Assuming a missile with weight M (lb), velocity v (ft/sec), area A (in.²)
and diameter D(in.), there are a large number of penetration correlations
available to relate the energy of the missile to the barrier needed to contain
it (*32*).

For *steel*, the usual correlation used is the Stanford equation (*33*)

$$E/D = (S/46,500)(16,000T^2 + 1500WT/W_s) \tag{5.32}$$

where E is the missile energy, S is the ultimate tensile strength of the steel
(psi), T is the barrier thickness (in.), W is the length of a steel square side
between rigid supports (in.), and W_s is the length of a standard width (4 in.).

For *concrete*, the Petry formula is commonly used (*34*)

$$d = (KM/A) \log_{10}(1 + v^2/215,000) \tag{5.33}$$

This gives the penetration depth d (in.) when K, an experimental coefficient,
is obtained for the particular case under consideration. However, the Petry
formula is rather optimistic compared to a large number of other formulas
that have been derived largely by military and naval research. A recommended correlation is

$$d = (282M/2(S_u)D^2) (D)^{0.215}(v \cdot 10^{-3})^{1.5} + (D/2) \tag{5.34}$$

where S_u is the compressive strength of concrete (psi) (*35*).

Some reactor systems have included specific missile shields built above
the vessel head. The Enrico Fermi reactor missile shield completely enclosed the head and included an aluminum absorber section to reduce the
missile impact energy. The component is illustrated in Fig. 5.18.

It should be noted that the most efficient missile barrier is distance and
therefore it may be possible to specify a minimum ceiling height for the
containment or any attendant hot cells so that penetration does not become
a problem. Such a possibility will depend largely on layout and refueling
method.

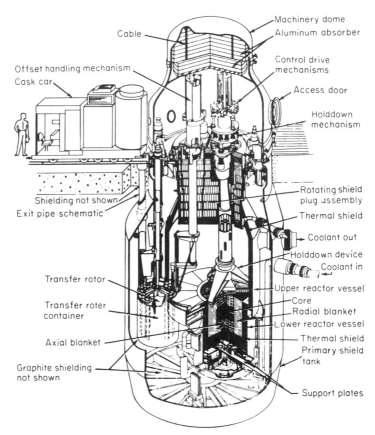

Fig. 5.18. Cutaway drawing of the Enrico Fermi reactor showing in particular the machinery dome and its aluminum absorber intended to act as a missile shield (*29*). (Courtesy of Atomic Power Development Associates, Inc.)

External missiles are treated in much the same way. Sources of missiles from such items as a runaway turbine, components of overflying planes, and tornado-driven debris are assessed for the energies involved and their penetrations are calculated by the same formulas as those given above. For very large and improbable external missiles, barriers cannot be provided but it is still necessary to know the probable consequences in order to ensure that the reactor fuel itself would not be damaged and that no radioactive fission products would be released to the atmosphere. Generally, the concern is not with the reactor itself, which is well protected by the massive head plug, but with the spent fuel storage pit which may only be covered by a thin lid, but which, nevertheless, contains a large inventory of fission products.

REFERENCES

1. "Atomic Radiation: Theory, Biological Hazards, Safety Measures, Treatment of Injury," 8th ed. RCA Service Co., Government Services, Camden, New Jersey, 1961.
2. N. A. Frigerio, "Your Body and Radiation." AEC information booklet. Oak Ridge National Laboratory, Oak Ridge, Tennessee, March 1969.
3a. R. O. McClellan, M. E. Kerr, and L. K. Bustad, Reproductive performance of female miniature swine ingesting strontium-90 daily. *Nature* **197**, 670 (1963).
3b. "Environmental Effects of Producing Electric Power." Hearings before Joint Comm. At. Energ., October and November 1969, Part I, U.S. Government Printing Office, Washington, D.C., 1969.
4. K. Z. Morgan, Acceptable risk concepts. Lecture. Pittsburgh Sect. *Amer. Nucl. Soc., Pittsburgh, Pennsylvania, November 18, 1969.*
5. J. Joseph, Medical miracle vs. atomic death. *True Magazine* **50**, 72, (1969).
6. J. H. Wright, Electrical power generation and the environment. *Westinghouse Eng.* **30** (3), 66 (1970).
7. Recommendations of the international commission on radiological protection (adopted September 17, 1965). ICRP Publ. 9. Pergamon, New York, 1966.
8. Code of Federal Regulations—Section 10 Atomic Energy Law Reports. Pt. 20: Standards for Protection against Radiation (January 1961), Pt. 50: Licensing of Production and Utilization Facilities (February 1969), and Pt. 100: Reactor Site Criteria (April 1962).
9. Based on paragraph 75 of Recommendations of the International Commission on Radiological Protection (adopted September 17, 1965). ICPR Publ. 9. Pergamon, New York, 1966.
10a. P. Spiegler, J. G. Morgan, M. A. Greenfield, and R. L. Koontz, Characterization of aerosols produced by sodium fires—Vol. I. NAA-SR-11997, Vol. I, Atomics International, Canoga Park, California. May 1967.
10b. M. A. Greenfield, R. L. Koontz, and D. H. Hausknecht, Characteristics of aerosols produced by sodium fires—Vol. II. AI-AEC-12878, Atomics International, Canoga Park, California. October 1969.
10c. A. W. Castleman, Jr., A summary of recent progress in aerosol research at Brookhaven National Laboratory. BNL-14271. Brookhaven Nat. Lab., Upton, New York, 1969.
11. M. Smith, ed., "Recommended Guide for the Prediction of the Dispersion of Airborne Effluents." ASME, New York, May 1968.
12. F. Pasquill, "Atmospheric Diffusion." Van Nostrand, Princeton, New Jersey, 1962.
13. "Post-Weld Heat Treatment," 1968 ed., Sect. VIII, ASME Boiler and Pressure Vessel Code, Unfired Pressure Vessels, Section B, Article 13, N 1342.
14a. "Guide to Nuclear Power Cost Evaluation," Vol. 2, "Land, Improvements, Buildings and Structures." H. J. Kaiser Co., Oakland, California, TID 7025, Vol. 2, March 1962.
14b. J. G. Yevick and A. Amarosi, eds., "Fast Reactor Technology: Plant Design." M.I.T. Press, Cambridge, Massachusetts, 1966.
15a. W. J. McCarthy, Jr., and W. H. Jens, A Review of the Fermi Reactor Fuel Damage Incident and a Preliminary Assessment of Its Significance to the Design and Operation of Sodium Cooled Fast Reactors, *Proc. Int. Conf. Safety of Fast Reactors, Aix-en-Provence, September 1967.* Papers Va-1-1—Va-1-23, Commissariat à l'Énergie Atomique, Paris, 1967.

15b. Final Summary Rept. on the Examination of Fermi Subassemblies M-140, M-127, M-098, and M-122 to PRDC, November 1968. Battelle Memorial Inst., Columbus, Ohio, 1968.

16. T. J. Thompson, Accidents and destructive tests. *In* "The Technology of Nuclear Reactor Safety," Vol. 1, "Reactor Physics and Control" (T. J. Thompson and J. G. Beckerley, eds.), p. 608, Sects. 3.2, 3.11. M.I.T. Press, Cambridge, Massachusetts, 1964.

17. L. Cave and T. F. Williams, Safety assessment of pressure-tube heavy-water reactors by probability methods. "Heavy Water Power Reactors," p. 831–863, Paper SM-99/19. IAEA, Vienna, 1968.

18. F. R. Farmer and E. V. Gilby, A method of assessing fast reactor safety. *Proc. Int. Conf. Safety of Fast Reactors, Aix-en-Provence, September 1967*, Paper VI-2-1. Commissariat à l'Énergie Atomique, Paris, 1967.

19. H. J. Teague, Cooling failure in a subassembly. "An Appreciation of Fast Reactor Safety (1970)," Chapter 2. Safeguards Div., A.H.S.B., UKAEA, 1970.

20. O. D. Kazachkovskii *et al.*, Investigation of a working fuel element bundle from the BR-5 reactor with plutonium dioxide fuel. *At. Energ.* **24** (2), 136–143 (1968).

21a. M. D. Carelli, H-D. Garkisch, and J. Graham, private communication, Westinghouse, 1970.

21b. J. Graham and J. Versteeg, Consequences of Molten fuel ejection in a LMFBR, in *Proc. Am. Nucl. Soc. Nat. Topical Meeting on New Developments in Reactor Mathematics and Applications, Idaho Falls, Idaho, 1971*, to be published.

22. J. F. Proctor, Explosion containment for nuclear reactor vessels. *Nucl. Safety* **7** (4), 459 (1966).

23. J. F. Proctor, Adequacy of explosion-response data in estimating reactor-vessel damage. *Nucl. Safety* **8** (6), 565 (1967).

24. R. H. Cole, "Underwater Explosions." Princeton Univ. Press, Princeton, New Jersey, 1948.

25. Final report of the SL-1 recovery operation. USAEC Rept. IDO-19311. General Elect. Co., Idaho Test Station, Idaho. July 27, 1962.

26. G. R. Warren and T. W. Rice, An assessment of reactor safety modelling techniques based on the accident to the SL-1 boiling water reactor. Brit. Rept. AWRE-R-3/65, A.W.R.E., Aldermaston, England. August 1965.

27a. Y. W. Chang, J. Gvildys, and S. H. Fistedis, Two dimensional hydrodynamics analysis for the primary containment. ANL 7498. Argonne Nat. Lab., Argonne, Illinois, June 1969.

27b. S. Veyo, Private communication, Westinghouse Research and Development, Pittsburgh, Pennsylvania. 1969.

28a. G. Cinelli, J. Gvildys, and S. H. Fistedis, Inelastic response of primary reactor containment to high-energy excursions. ANL 7499. Argonne Nat. Lab., Argonne, Illinois, August 1969.

28b. Argonne National Laboratory Reactor Development Progress Report, Nuclear Safety Research and Development, p. 149, ANL-7679, March 1970.

29. Enrico Fermi atomic power plant technical information and hazards summary report, Vols. 1, 2, and 3. Power Reactor Develop. Co., Monroe, Michigan. October 1962.

30. Fast reactor descriptive manual. IG-Rept.-170 (D), Vol. 1, Sect. A.2. 1958.

31a. G. Jansen, Fast reactor fuel interaction with a concrete floor after a hypothetical core melt-down. BNWL-CC-2369, Pacific Northwest Lab., Richland, Washington. January 1970.

31b. T. H. Row, Interactions of molten UO_2 and refractory materials. ORNL-TM-1828. Oak Ridge Nat. Lab., Oak Ridge, Tennessee, June 1967.

32. R. C. Gwaltney, Missile generation and protection in light-water cooled power reactor plants. ORNL-NSIC-22. Oak Ridge Nat. Lab., Oak Ridge, Tennessee, September 1968.

33. R. W. White and N. B. Botsford, Containment of fragments from a runaway reactor. Rept. SRIA-113. Stanford Res. Inst., Stanford, California, September 15, 1963.

34. C. R. Russell, "Reactor Safeguards." Macmillan, New York, 1962.

35. Effects of impacts and explosions. Summary Tech. Rept. of Div. 2, Vol. 1. Nat. Defense Res. Comm., Washington, D.C., 1946.

Note added in proof: The AEC has, at the time of printing, formulated much lower numbers to define its new "as low as practicable" criteria for power reactor radioactive emissions. They lower the maximum body dose to anyone outside a plant's boundary by a factor of 100 to 5 mrem and also place restrictions on exposure to the total population within a 50-mile radius to 100 man-rems. The man-rem is an integrated exposure unit currently in use in Britain.

CHAPTER 6

LICENSING

The ultimate objective of the safety engineer is to be able to *show* that the system which he has nurtured *is* indeed safe, and to be able to show this with such conviction as to obtain licenses to build and to operate the plant. To this end he provides a safety evaluation of the plant as a report in support of the license application.

6.1 Codes, Standards, and Criteria

The safety evaluation will first show that the system design and its operation adhere to certain codes, standards, and criteria that have been laid down to guide safe design.

These guidelines can be defined as rules of conduct: the criteria to provide general design aims for the system; standards to provide design limits and values to be used in the design methods laid down by codes.

6.1.1 CODES

Codes are produced by engineering societies, both as an aid to their members and as a public service. Codes are not government controlled even though they are sometimes the subject of joint effort or subsequent agreement by the government agencies. The American Society of Mechanical Engineers (ASME) codes are an example. Among other things they provide methods for nuclear vessel design (*1*). Figure 6.1 shows an example of a specific code for jointing sections of wall of unequal thickness from Section III of the ASME code for nuclear vessels. The Institution of Electronic and Electrical Engineers (IEEE) codes, on the other hand, deal with electrical components and systems (*2a,b,c*).

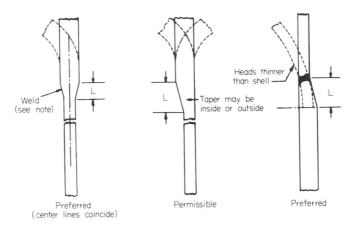

Fig. 6.1. N-466 category A and B joints between sections of unequal thickness. ASME code: section III, nuclear vessels (*1*). The length of required taper *L* may include the width of the weld. In all cases *L* shall not be less than 3 times the offset between the abutting plates.

Section 10 of the Code of Federal Regulations (CFR) for Atomic Energy (*3*) is important in the sense that these are the codes by which licensing bodies judge the applicants. The codes 10 CFR 20 and 10 CFR 100 for radiological evaluation have been referred to previously. Section 6.3 refers to 10 CFR 50 as a guide to safety evaluation report preparation.

6.1.2 STANDARDS

Standards are also produced by the engineering societies and governmental institutions to provide limits for the design of various components.

For example, the American Society for Testing Materials (ASTM) provides for piping standards (*4*), and the United States of America Standards Institution (USASI) provides a vast number of everyday standards as well as those of use in the nuclear industry (*5*).

6.1.3 CRITERIA

General design criteria are design rules (*6*) that could be expanded to include codes and standards as *specific* criteria. Chapter 3 has already dealt with criteria in some detail.

To show how *specific* criteria guide the designer without telling him either how to do the analysis or what design limits to use (apart from the ultimate ones), consider *General* Criterion 50 (*6*) rewritten very slightly to apply to fast reactor systems.

50: *Containment Design Basis* (for Fast Reactors)

The reactor containment structure, including access openings, penetrations, and containment heat removal systems shall be designed so that the containment structure and its internal components can accommodate, without exceeding the design leakage rate and with sufficient margin, the calculated pressure and temperature conditions resulting from the maximum credible accidental energy release. This margin shall reflect considerations of (1) the effects of potential energy sources which have not been included in the determination of peak conditions, such as the energy in steam generators and energy from chemical reactions which may result from degraded emergency cooling functioning, (2) the limited experience and experimental data available for defining accident phenomena and containment responses, and (3) the conservatism of the calculational model and input parameters.

The corresponding *specific* criteria for a two-compartment containment building might be:

Containment Criteria

50.1 Plant containment will be provided to limit radioactivity release to the environment. This release will not exceed the limits described in 10 CFR 20 for normal operations and for transient situations which might be expected to occur, and 10 CFR 100 for potential accidents which have a very low probability of occurrence.

50.2 This inner containment structure will be designed to withstand peak pressures and temperatures above those which could occur as a result of the design basis accident, taking into account the calculational uncertainties. The structure will be designed to P_1 psig excess pressure associated with a maximum temperature of $T_1°F$ at the liner, but not necessarily simultaneously.

50.3 The design basis accident for the inner containment will be the worst credible core disruptive accident.

50.4   The reactor inner containment structure atmosphere will be at all times inerted with a maximum oxygen content of 2%. A sodium fire will be deemed to be incredible within the reactor containment structure.

50.5   The reactor inner containment structure will be designed to have a leakage rate of less than $L_1 v/0$ per day† even under the design peak pressures and temperatures.

50.6   The outer containment building will be designed to withstand peak pressures and temperatures above those which could occur as a result of its design basis accident, taking into account the calculational uncertainties. The building will be designed to P_2 psig excess pressure associated with a surface temperature of T_2°F.

50.7   The design basis sodium fire for this outer containment building will be that associated with a sodium pool in an open IXH vessel.

50.8   The outer containment building will be designed to have a leakage rate of less than $L_2 v/0$ per day even under the design peak pressures and temperatures.

50.9   The design will preclude open interconnection between the atmosphere of the inner containment structure and the atmosphere of the outer containment building unless the sodium temperature in the primary circuit has been reduced to below 300°F.

50.10   Primary coolant components of the IHXs and pumps will be drained of primary coolant before the components are removed to maintenance areas.

50.11   The design of both containment buildings will accommodate earthquake load-ings and other environmental loadings that the site might make necessary, in addition to normal static loads.

The design pressures and temperatures P_1, P_2, T_1, T_2 and leakages L_1, L_2 are set with suitable margins from the safety evaluation of certain design basis accidents as outlined in Section 5.4.

Within these still fairly wide rules the designers' codes and standards must set the methods and define the numerical limitations within which he works to meet the criteria. There may be a number of design methods of meeting the leakage limitations set by 50.5 for example, even at the condi-tions specified in 50.2 to restrict the choice. Ultimately there are other criteria that also come into play in the final choice of solution, the most important being the cost. The designer has many masters to satisfy, and he uses the criteria as guide lines in his work.

6.2 Regulatory Processes

The regulatory processes following an application for a construction or an operating license involve judges, juries, witnesses, plaintiffs, and the respective counselors and it is important to get the interrelationship of the various bodies into perspective.

---

† Note: this notation indicates volume percent per day, abbreviated throughout this volume as vol%/day.

6.2.1 Bodies Concerned with Safety for Nuclear Power Plants in the USA

6.2.1.1 *Atomic Energy Commission*

The AEC is involved at many stages of plant design and licensing through its Division of Reactor Development and Technology and through its Division of Reactor Licensing and the Advisory Committee on Reactor Safeguards.

(a) Division of Reactor Development and Technology (RDT). This division deals with the technical side of fast reactor technology rather than with licensing. It places contracts for all aspects of safety research with industry and the National Laboratories. It is concerned with the promotion of Nuclear Power. It is also available as an expert witness during licensing.

(b) Division of Reactor Licensing (DRL). This division, which deals with standards, licensing and compliance, checks the safety submission for formal compliance with safety criteria.

(c) Advisory Committee on Reactor Safeguards (ACRS). This committee makes the prime decision on the safety of the nuclear power plant. It receives the presentation of the safety evaluation from the applicants and after advice from DRL makes recommendations to the AEC Commissioners.

Figure 6.2 shows the relationship between the AEC and these bodies

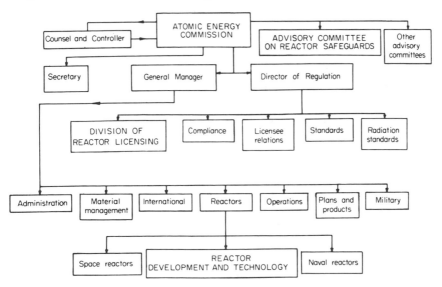

Fig. 6.2. United States Atomic Energy Commission organization (1971).

in 1971. It has been suggested that in the future the licensing process (DRL) should be separated from the technical promotional side of the AEC work (RDT).

The ACRS is composed of a set of independent non-AEC experts from a very wide range of disciplines. In 1971 they included: a professor of nuclear engineering, a physicist at Brookhaven National Laboratory, a consultant in metallurgy at Battelle Memorial Institute, a consultant in mechanical reactor engineering, a professor of chemical engineering, a consultant in hydraulic engineering and lake biology, the chairman of the board of Crown Central Petroleum Corporation, a senior engineer at Argonne National Laboratory, a consultant in industrial chemistry, a professor of civil engineering, a consultant in chemical engineering, a physicist at Los Alamos Scientific Laboratory, a director of the Sanitary Engineering Research and Radiological Research Laboratory, a professor of nuclear engineering, and a senior physicist at Argonne National Laboratory.

The chairman of the committee is changed annually from among the committee members. The 14 members encompass a wide range of talents, and it is clear that such a committee makeup is liable to produce a very balanced outlook of the overall safety of any given nuclear plant under consideration.

Within RDT there is a technical organization led by a director (Milton Shaw for the year 1971) and assistant directors in charge of reactor engineering, nuclear safety, project management, reactor technology, plant engineering, program analysis, engineering standards, and army reactors.

Within the nuclear safety department there are groups working on research and development, engineering and tests, analysis and evaluation, and environment and sanitary engineering, although almost all departments have interests within the safety field.

RDT in practice has a dual function. As the technical advisory arm of the AEC, the division will enter the licensing process as the AEC expert advisor and, therefore, the division must keep abreast of technical developments in all the relevant fields. RDT also places contracts for safety research and development with national laboratories and industry and the division is therefore in a position to direct national research and development to a large extent as it administers the available funds. It bears great responsibility for the direction of the LMFBR program.

6.2.1.2 *National Laboratories*

The National Laboratories are run under contract by private bodies: Argonne National Laboratory (ANL), run by University of Chicago; Los

Alamos Scientific Laboratory (LASL), run by University of California; Oak Ridge National Laboratory (ORNL), run by Union Carbide; Brookhaven National Laboratory; and Savannah River National Laboratory.

They are almost wholly financed by the AEC and indirectly organized by the AEC. Their main function is to engage in basic research and code development, using facilities which are too large and expensive for private industry to operate.

Argonne National Laboratory and its facilities in the National Reactor Testing Station (NRTS), Idaho, forms the major force in fast reactor development. A good deal of experimental work including EBR-II and the TREAT facility comes within the ANL orbit. The ANL code center acts as a clearing house for ANL and industry codes developed under AEC contract, available for the good of fast reactor development.

Oak Ridge National Laboratory, besides having an experimental fast reactor program devoted to both sodium facilities and gas-cooled work, runs the National Safety Information Center (NSIC). The NSIC distributes abstracts on the user's selected safety topic to the user on a regular basis. The service, operated as a totally computerized system, is free to any

```
ACCESSION 47059                    992-01      B                09-1--70

GAST K + KELLER K + PEPPLER W + SCHIKARSKI W + SCHLECTENDAHL EG +
SCHULTHEISS GF + WILD H
REPORT ON THE SAFETY ANALYSIS OF THE FAST SODIUM-COOLED REACTORS
KERNFORSCHUNGSZENTRUM, KARLSRUHE (WEST GERMANY)
NP-18150 +.  37 PAGES, AUGUST 1969 (IN GERMAN)

FAILURE PROPAGATION AND ACCIDENTS IN FAST, SODIUM-COOLED FUEL ELEMENTS
ARE DISCUSSED. METHODS BY WHICH FAILURE PROPAGATION OCCURS ARE
CONSIDERED - EFFECTS OF THE SODIUM COOLANT BOILING IN CORE ARE
ANALYSED.  SODIUM BOILING EXPERIMENTS FOR DETERMINING THERMODYNAMIC
AND NEUTRONICS BEHAVIOR ARE DESCRIBED.  RADIOACTIVITY RELEASE UNDER
ACCIDENT CONDITIONS IN A SODIUM-COOLED FAST REACTOR IS EVALUATED.
AEROSOL AND PARTICULATE GENERATION AND TRANSPORT UNDER ACCIDENT
CONDITIONS ARE ANALYSED.  INCIPIENT AND POOL BOILING MECHANISMS ARE
DISCUSSED.

AVAILABILITY - CLEARINGHOUSE FOR FEDERAL SCIENTIFIC AND TECHNICAL
INFORMATION, SPRINGFIELD, VA.  22151  $3.00 COPY, $0.65 MICROFICHE

*REACTOR, FAST + REACTOR, LMCR + SODIUM + *SAFETY ANALYSIS + VAPOR,
VAPOR PRESSURE + FAILURE MODE ANALYSIS + BOILING + REACTOR KINETICS +
RADIOACTIVITY RELEASE + AEROSOL + HEAT TRANSFER
```

Fig. 6.3. National Safety Information Center abstract card.

engineer engaged in safety work. It is very comprehensive. Figure 6.3 shows an example of an abstract card produced by the NSIC and Fig. 6.4 shows an example of a page from an NSIC-produced listing of articles which have appeared in the journal *Nuclear Safety*, also issued by ORNL [see general references, Chapter 1].

08-5-2-455   CABRI - A TEST REACTOR FOR SAFETY STUDIES
             MILLOT, J. P.
             CADARACHE NUCLEAR RESEARCH CENTER, FRANCE
                  THE SAFETY OF SWIMMING-POOL REACTORS HAS BEEN THOROUGHLY
                  STUDIED IN FRANCE. THE CABRI REACTOR, DESIGNED FOR THAT
                  PURPOSE, WAS INITIALLY USED TO INVESTIGATE REACTIVITY
                  ACCIDENTS. LOSS-OF-COOLANT-FLOW ACCIDENTS WERE ALSO STUDIED,
                  AND THIS WORK IS BEING EXTENDED WITH A NEW FACILITY THAT BECAME
                  OPERATIONAL THIS YEAR, CABRI PUISSANCE. A PROGRAM OF FUEL
                  TESTING FOR BOTH WATER AND FAST REACTORS, INVOLVING CABRI AS A
                  CAPSULE-DRIVER CORE, IS BEING ESTABLISHED.

08-5-2-461   RADIATION DAMAGE TO PRESSURE VESSEL STEELS
             WECHSLER, M. S.
             OAK RIDGE NATIONAL LABORATORY, OAK RIDGE, TENNESSEE
                  THE EFFECTS OF NEUTRON IRRADIATION ON THE PROPERTIES OF
                  STRUCTURAL MATERIALS ARE OF SIGNIFICANT INTEREST TO THE REACTOR
                  INDUSTRY BECAUSE OF THEIR POSSIBLE RELATION TO THE INTEGRITY OF
                  REACTOR PRESSURE VESSELS. IRRADIATION EFFECTS ON
                  PRESSURE-VESSEL STEELS ARE CONSIDERED. THE VARIABLES THAT
                  AFFECT THE NIL CONDUCTIVITY TRANSITION TEMPERATURE ARE
                  SEPARATED INTO TWO CATEGORIES - MATERIALS VARIABLES AND
                  RADIATION VARIABLES. AMONG THE MATERIALS VARIABLES CONSIDERED
                  ARE CHEMICAL IMPURITIES (SUCH AS BORON, CARBON, AND NITROGEN),
                  GRAIN SIZE, AND METALLURGICAL STRUCTURE ASSOCIATED WITH
                  WELDING. THE RADIATION VARIABLES INCLUDE DOSE, DOSE RATE,
                  NEUTRON SPECTRA, AND IRRADIATION TEMPERATURE. OTHER AREAS
                  REVIEWED INCLUDE POSSIBLE APPLICATIONS (AND LIMITATIONS) OF
                  FRACTURE MECHANICS TO THE PROBLEM OF RADIATION EMBRITTLEMENT IN
                  PRESSURE-VESSEL STEELS AND THE EFFECTS OF IRRADIATION ON THE
                  FATIGUE STRENGTH OF STEELS.

08-5-3-470   RELIABILITY ANALYSIS OF ENGINEERED SAFEGUARDS
             GARRICK, B. J. + GEKLER, W. C.
             HOLMES AND NARVER, INC., LOS ANGELES, CALIFORNIA
                  THE REQUIREMENTS FOR ASSESSING RELIABILITY IN ENGINEERED
                  SAFEGUARDS ARE REVIEWED, AN EXAMPLE OF RELIABILITY ANALYSIS IN
                  A TYPICAL EMERGENCY CORE-COOLING SYSTEM IS GIVEN, AND A PLAN IS
                  PRESENTED FOR USE OF RELIABILITY TECHNIQUES IN NUCLEAR SYSTEMS
                  SAFETY ANALYSIS. RELIABILITY CAN NOW BE USED IN MANY WAYS IN
                  NUCLEAR SAFETY, BUT, BEFORE IT CAN BE A TRULY EFFECTIVE TOOL,
                  CERTAIN ADDITIONAL REQUIREMENTS MUST BE MET. RELEVANT
                  PROBABILISTIC MODELS MUST BE DEVELOPED, AND REALISTIC
                  STATISTICAL DATA ARE NEEDED. TECHNIQUES EXIST FOR DEVELOPING
                  RELEVANT MODELS. STATISTICAL DATA NOW DERIVE LARGELY FROM
                  OPERATING EXPERIENCE. THESE DATA, ALTHOUGH OF SOME USE, MUST BE
                  IMPROVED TO PERMIT CONFIDENCE IN MATHMATICAL EVALUATION OF
                  RELIABILITY. THE IMPORTANCE OF RELIABILITY LIES NOT SO MUCH IN
                  NUMBERS GENERATED AS IT DOES IN REQUIRING SYSTEMATIC GUIDANCE
                  OF SAFETY ANALYSIS AND TESTING AND ACCIDENT-PREVENTION
                  PROGRAMS.

08-5-3-479   RELIABILITY ESTIMATES AND REACTOR SAFETY SYSTEM OPERATION
             SCOTT, R. L.
             OAK RIDGE NATIONAL LABORATORY, OAK RIDGE, TENNESSEE
                  TWO PAPERS PRESENTED AT THE WINTER 1966 MEETING OF THE AMERICAN
                  NUCLEAR SOCIETY ARE SUMMARIZED. ONE PAPER DISCUSSES DATA
                  OBTAINED FROM FIVE POWER-REACTOR OPERATING AGENCIES IN ORDER TO
                  MAKE RELIABILITY ESTIMATES. SINCE THE DATA WERE OBTAINED FOR
                  PURPOSES OTHER THAN RELIABILITY PREDICTION, IT HAS LIMITED
                  APPLICABILITY. RELIABILITY ESTIMATES ARE PRESENTED FOR
                  EMERGENCY CORE-COOLING SYSTEMS, EMERGENCY POWER SYSTEMS, AND
                  SECONDARY SHUTDOWN SYSTEMS. THE OTHER PAPER DISCUSSES
                  REACTOR-SAFETY SYSTEM OPERATION DATA OBTAINED FROM THE SAME
                  FIVE POWER-GENERATING PLANTS. THE COMPOSITE DATA INDICATE THAT
                  REAL AND SPURIOUS SCRAM RATES ARE EACH BETWEEN FIVE AND SIX PER
                  YEAR. MOST OF THESE SCRAMS OCCURRED AT LOW REACTOR POWER AND
                  PRODUCED ONLY ONE OR TWO POWER-GENERATION OUTAGES PER YEAR.

Fig. 6.4. Page of National Safety Information Center listing of articles which have appeared in *Nuclear Safety*.

References and abstracts are located by certain key words. Thus, by specifying a number of key words such as *reactor, fast, melt-down, accident-analysis*, the user is able to specify a very well defined topic of his interest on which he would like information.

In addition, ORNL produces survey documents on selected technical topics such as missile protection, nuclear reactors and earthquakes, etc. This service is also an AEC-financed information service.

6.2.1.3 *Privately Endowed Laboratories*

Certain other laboratories also deeply concerned with fast reactor technology development and safety are the Pacific Northwest Laboratory (PNL), run by the Battelle Memorial Institute; the Liquid Metal Engineering Center (LMEC), run by Atomics International Division of North American Rockwell, Inc.; and the Hanford Engineering Development Laboratory (HEDL), run by Westinghouse Electric Corporation.

These laboratories are also subject to AEC contracts and under commitment to the AEC for certain projects. HEDL's Westinghouse management is now engaged in bringing the FFTF facility into being, while LMEC provides sodium technology information and acts as a clearing house for failure data.

6.2.1.4 *Industry*

Three industrial giants are in current competition in the LMFBR field: Westinghouse Electric Corporation, General Electric Company, and Atomics International Division of North American Rockwell, Inc. Others such as Combustion Engineering, and Babcock and Wilcox are showing partial interest in the LMFBR market as it develops. All industrial effort is subject to AEC contracts and to utility support during this development time.

Finally there are the hundreds of utility companies who have a stake in nuclear power and therefore support one or another of the major industrial designers in the business. The utility would eventually be the owner of any fast-reactor power plant supplied by an industrial vendor. Therefore, the utility would be the applicant in seeking a license to build and operate.

6.2.2 LICENSING APPLICATION

There is a sequence of events in the licensing process that must take place before a license is granted.

(a) The application for the license to build is submitted by the utility as applicant, using technical information supplied by the industrial designer. This technical information includes a preliminary safety analysis report, the PSAR.

(b) The DRL receives the application and gives a copy to the ACRS. It puts a copy on view in the public room to give objectors access to all the information. It considers the application and the evaluation report and then submits its report, also to the ACRS.

(c) The ACRS considers the application, the safety evaluation, and the DRL report. It then reports in turn to the AEC commissioners who make the official ruling. However, the ACRS by this time has made a recommendation which carries considerable weight. The AEC Commissioners are

TABLE 6.1

AEC QUESTION 14.1 ON INDIAN POINT 2

14.1 Calculate the required iodine reduction factor necessary to meet the 10 CFR 100 guideline values for the available exclusion and low population zone radii using the following assumptions:

(a) Power level of 3216 MWt.

(b) TID-14844 fission product release fractions (100% of the noble gases, 50% of the iodines, 1% of the solids).

(c) An iodine plateout factor of 2.

(d) 10% of the airborne iodines being methyliodide.

(e) Containment leak rate of 0.1%/day for the first 24 hr, 0.045%/day thereafter.

(f) With the following meteorology:

(1) Pasquill Type F, 1 m/sec, nonvarying wind direction, and volumetric building wake correction factor with $C = \frac{1}{2}$ and the cross-sectional area of the containment structure for the first 8 hr.

(2) From 8 to 25 hr, Pasquill Type F, 1 m/sec with plume meander in a 22.5° sector.

(3) From 1 to 4 days, Pasquill Type F and 2 m/sec with a frequency of 60%, Pasquill Type D and 3 m/sec with a frequency of 40%, with a meander in the same 22.5° sector.

(4) From 4 to 30 days, Pasquill Type C, D, and F each occurring 33.3% of the time with wind speeds of 3 m/sec, 3 m/sec, and 2 m/sec, respectively, with meander in the same 22.5° sector 33.3% of the time.

(g) A breathing rate of 3.47×10^{-4} m³/sec for the first 8 hr, 1.75×10^{-4} m³/sec from 8 to 24 hr, and 2.32×10^{-4} m³/sec thereafter.

unlikely to oppose this recommendation. During the ACRS review there is considerable communication between the committee and the utility–industrial designer partnership. The ACRS is likely to ask a series of searching questions before it is satisfied to make a recommendation to the commissioners. Table 6.1 details a typical question asked by the AEC, in this case during a PWR licensing application.

(d) The commissioners call a public meeting for public review and objection by groups or individuals if required. The licensing board at this meeting again reports to the commission on objections received.

(e) The commission then decides on the application and issues a construction license if they are satisfied that the plant is safe, and that public objections have been met, where these objections are relevant.

(f) After construction of the plant, the utility must again make an application, this time for an operating license. The procedure to obtain it is identical with steps (a)–(e) and again occupies many months. This second application is accompanied by another safety analysis document, a final one (FSAR), which incorporates all the latest work and includes answers to all the ACRS questions.

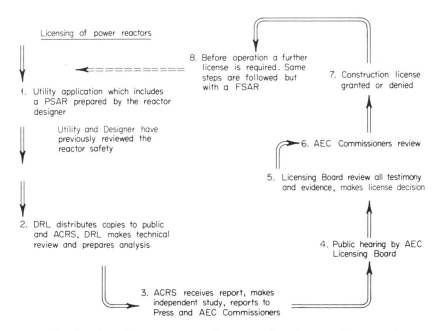

Fig. 6.5. Licensing of power reactors: how central-station atomic power plants are licensed and regulated (*7a*).

To summarize the procedure (Fig. 6.5):

1. application by utility (industrial vendor acts as an advisor);
2. DRL distribution of application and consideration of PSAR;
3. report by DRL to ACRS;   ⎱ questioning
4. report by ACRS to AEC commissioners;   ⎰ period
5. public review meeting;
6. report by licensing board to AEC commissioners;
7. construction license granted or refused.

Then after construction of the plant, there is a repetition of steps 1–6, following which the operating license may be granted or refused or simply delayed until all outstanding queries are settled.

Steps 1–7 occupy approximately eighteen months and there is approximately a two-year construction delay before the next application is made. During this time, the safety evaluation will be strengthened and made more specific to the design which will become fixed during the early construction period.

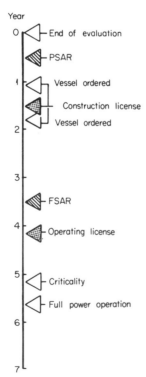

Fig. 6.6. Licensing time scale for a nuclear power plant.

6.2.3 TIME SCALE

The projected time scale for a 800 MWt fast reactor plant might occupy five years as shown in Fig. 6.6. It is worth noting that long term items such as the nuclear vessel and head must be ordered very soon after the PSAR is submitted. These order dates become the critical dates at which the design needs to be fixed to some extent.

The time scale for obtaining a license is very long: this is a public safety factor to allow the most searching reviews by the designers, the utility, the ACRS, and the public. Here again, a balance must be struck between a lack of critical review and an overemphasis of the review period with review for the sake of review. Both extremes can be detrimental to the plant safety.

6.3 Safety Analysis Reporting

During the licensing process at the times of the construction and operating license applications, evaluations of the reactor system safety must be submitted for consideration. They are called the Preliminary Safety Analysis Report (PSAR) and the Final Safety Analysis Report (FSAR).

There is approximately a three-year gap between them so that the analysis can be revised and updated if required (see Fig. 6.6). The national research and development safety program is so organized that there will be new information constantly appearing, which may enable the plant to be better optimized rather than to have it overemphasize safety.

The production of these safety analysis reports is standardized according to 10 CFR 50 and some interpretive documents (7b). This standardization allows the licensing authorities to relate safety evaluations to each other and to check rapidly whether all the required information is present. The safety analysis document is divided into separate parts:

Section	I.	Introduction and summary
	II.	Site
	III.	Reactor
	IV.	Reactor coolant system
	V.	Containment system
	VI.	Engineered safeguards
	VII.	Instrumentation and control
	VIII.	Electrical systems
	IX.	Auxiliary and emergency systems

X.   Steam and power conversion systems
XI.   Radioactive wastes
XII.   Conduct of operations
XIII.   Initial tests and operation
XIV.   Safety analysis
XV.   Technical specifications

Although this format is not strictly adhered to by all applicants, the formats of most PSARs are approximately the same. It is, however, possible that this format may vary in the future with the advent of a standardized SAR with even standard sections and wordings.

6.3.1 SECTIONS I–XIII OF THE SAR

These comprise a descriptive outline of the whole plant, together with specific reference to the safety of each system. Compliance with the relevant safety criteria is demonstrated in these sections and the codes and standards used are noted.

As an example, Section VIII (electrical systems) requires comment on the following subjects: design basis—electrical system; design—network interconnections, station distribution system, emergency power, and tests and inspections. Section III (reactor) requires discussion of design bases—nuclear, reactivity, mechanical, and thermal hydraulic; design—nuclear characteristics and evaluation, mechanical characteristics and evaluation, thermal hydraulic characteristics and evaluation; and safety limits and conditions—tests and inspections.

6.3.2 SECTION XIV ON SAFETY ANALYSIS

The safety analysis presentation is the purpose of the PSAR document. It requires a core and primary circuit analysis and a standby safeguards analysis with emphasis in each section on abnormalities considered, identification of causes, and a complete analysis with methods and results.

Table 6.2 is an outline of a safety analysis section, as an example of exactly what might be covered in such an analysis, in this case for a three-loop LMFBR.

This document would be prepared by the design and safety groups of the industrial vendor for their utility customer and the utility would submit it to the AEC in support of their application. The vendor acts as an expert witness for the utility if any questions on the work arise.

TABLE 6.2

TABLE 6.2 (*continued*)

6.4 Other Siting Considerations

The regulatory process includes a public hearing and public groups and individuals are allowed to make comment on the proposed power plant and the applicants' submissions. The attitude of the public is influenced by the safety of the plant and radiological limitations which are set by possible normal and off-normal emissions which would not be expected.

However the public are also influenced by aesthetic and ecological considerations as well as by possible radiological consequences. It is therefore pertinent to make some reference to these considerations before leaving the subject of license application. To put these effects into context, the fast breeder will be compared to thermal reactors and fossil-fueled plants.

6.4.1 AESTHETIC EFFECTS

Such disturbances may be separated into visual and political considerations.

6.4.1.1 *Visual*

Individuals living in the locality, preservation groups, and the power industry itself are all concerned in lessening the impact of the power plant on the countryside. This can be done by siting in relation to geographical features, architectural design, and landscaping.

The power plant may be sited advantageously to diminish the height of buildings, to hide transformer yards, to remove power transmission lines from skylines, and to use existing roads and facilities where possible. Architectural design enters into the picture by the design, placing, and coloring of buildings that blend rather than strike the eye. The British Steam Generating Heavy Water Reactor (7c), for example, was built in the lee of a small hill, half below grade; the building was colored green and oriented to diminish its effect on the area, all in order to improve its visual impact. Landscaping of every industrial installation is of course commonplace although less effective with very large power plant buildings.

Nuclear power plants are clearly much simpler to make visually attractive than fossil-fueled power plants because of the lack of a smoke stack and the lack of vast fuel storage areas around the site. Both types of system have the same problems of power transmission and their associated high voltage lines.

The fast reactor is a smaller reactor for a given power output, but as far as associated plant is concerned, it is essentially the same size as the thermal light water systems and has the same siting visual effects as other nuclear plants.

6.4.1.2 *Political*

The siting of a power plant in a particular area will have an impact on the industrial growth of that area in terms of power availability in the long run and in terms of local employment in the short run. Both of these effects are political in nature, and both have positive and negative effects depending on whether the area was originally rural or industrial and whether it was a thriving or depressed (in terms of lack of employment) locality. Political effects are largely power plant independent, being associated largely with the area and its population. However, they should not be forgotten.

6.4.2 THERMAL EFFECTS

All power plants have excess and unusable heat derived from the use of steam turbines to generate electricity. This excess heat must be rejected from the system in one of several ways: by sea water cooling, by river cooling, by pond or lake cooling, by cooling towers to the atmosphere, or by using the heat for other applications. All these methods have been used by different power plants.

The concern here is that the rejected heat not be allowed to alter thermal conditions in the sea, the river, or the atmosphere, and cause detrimental consequences to the ecology. Suggested effects have included the disturbance of oyster beds, an upset of river balance so that fine game fish leave and coarse fish remain, and even local changes in climate. Unfortunately, detrimental effects, true or false, are more newsworthy than positive effects.

6.4.2.1 *Power Plant Heat Generation*[†]

Table 6.3 lists the power conversion efficiencies for nuclear and fossil-fueled plants. It can be seen that something over 60% of the heat produced within a power plant is in excess of that which can be used in creating electricity. This inefficiency is suffered by all heat machines and the thermal nuclear plants, at present using lower temperature steam, are somewhat more inefficient than the fossil-fueled systems. However, the difference is

[†] See Wright (8).

TABLE 6.3

POWER CONVERSION EFFICIENCIES[a]

Heat machine	Steam conditions	Percentage efficiency (%)
Fossil fueled steam electric	1000°F at 1800–3500 psia	37–38
Thermal nuclear plants	600°F at 800–1000 psia	31–33[b]
Gas turbines	—	approx. 20
Automobile engines	—	approx. 10

[a] See Wright (*8*).
[b] With the 1000 MWe size LMFBR's, this efficiency can be increased to the neighborhood of 41 or 42%.

not significant and, size for size, the thermal effects of these competing systems are the same in magnitude. Fast reactors will eventually be capable of higher efficiencies in the large sizes. However, about 25% of the excess heat from a fossil-fueled plant is discharged directly from its stack along with gaseous pollutants; therefore the amount of excess heat to be removed by cooling water is relatively less than for the light-water-cooled nuclear plants. With fast reactors, however, the amount of heat released to the cooling water by the nuclear and fossil plants will be much the same (*9a*).

To put the excess heat position in perspective, by the year 2000, approximately 1250 billion gallons of water per day will be required to remove excess heat. About 30% of this will be discharged to the sea and the remainder will require surface runoff water in some form for cooling. Since the total average daily runoff for the United States is about 1200 billion gallons, over 70% of this would be required to cool the power plants in existence at that time, or at least to transfer the excess heat to the ultimate heat sink of the atmosphere (*9b*).

6.4.2.2 *Heat Dissipation*

Salt water cooling has already been used for the large graphite-moderated thermal reactors of Britain which are situated mainly on coastal sites. Good condensing temperatures are obtained but the corrosion problems make the use of salt water more expensive than fresh water.

It has been suggested that the very slight temperature rise of the sea which results could be put to good use in keeping northern channels free from ice, in lengthening the lobster season in Maine, in allowing fish herding in Long Island Sound, or even in increasing the numbers of tropical fish in the south. Additional uses include warming recreational waters (*10*). However, adverse effects have been suggested that change existing fish populations. Thus limits are placed on temperature changes by the Federal Water Pollution Control Administration. In coastal or estuarine waters, the discharge of heated waste should raise the maximum daily temperature on a monthly mean basis no more than 4°F during fall, winter, and spring, and no more than 1.5°F during the summer.

Once-through cooling from rivers is the most economical cooling method and was the usual method applied in the past. For large plants, depending on the size of the river, temperature changes could be fairly large. The Vermont Yankee based on the Connecticut river might have raised the temperature at the river discharge by about 20°F (*10*). Such a change of river temperature would change the present ecological balance and result in a new balance with different fish and different plant groups. Such a change would most likely be an undesirable one although research is not complete on the exact effect that would be obtained (*9a*), and the undesirability has not yet been demonstrated. The limitations now set on rises in river temperatures are set by each state and they range below 10°F depending on the value of the temperature at any time (*11*).

Fresh water circulated through a power plant's condensers from cooling ponds or lakes with subsequent ejection to the atmosphere by radiation and convection is attractive, since the water is used over and over again in a self-contained system. The method appears to be economical.

The rejection of heat for other applications such as heating apartment buildings or even whole towns (*10*) has been practiced in local areas, but the method cannot be considered an overall solution for the rejection of excess heat, especially since many power plants are situated in relatively remote areas.

There are different types of cooling towers. The main division is between the wet evaporative kind, in which some of the cooling water is evaporated to remove heat, and the dry kind, in which a tubed radiator transfers heat. Both kinds can employ natural draft cooling or induced draft cooling. They are designed broadly to lower the temperature of the cooling water by about 15–17°F. Which kind is used in a particular application depends on the natural climate and its effect on the possibility of natural draft, the surrounding countryside, and the effect of physically obtrusive structures, and whether or not it is necessary to avoid the use of noisy blowers for induced draft versions.

The advantage of cooling towers is mainly that they are a method of cooling recirculating water by direct contact with the environment rather than through the medium of a river. An additional advantage is that although they are very tall, cooling towers occupy relatively little site area when compared to cooling ponds which may occupy hundreds of acres of land area.

In siting any reactor plant, including a fast reactor installation, a comprehensive assessment of the impact of cooling upon the environment is now a necessity and the AEC, as the licensing authority, now has the power to refer applications for plant licenses to agencies having legal jurisdiction in environmental matters and to require that the licensee observe certain applicable environmental limits. This extension of the AEC powers was enacted in 1969 (*12*).

6.4.3 AIR POLLUTION

The nuclear power plant is seen as an answer to the air pollution problem, since in operation its only emissions would comprise very low radioactive waste gases with a radioactive level comparable to natural background activity. The nuclear power plant has none of the chemical emissions which may be released in large quantities from a fossil-fueled plant.

The matter of radioactive emission from an operating power plant is dealt with in Chapter 5. It should also be noted that modern nuclear plants are now being designed for essentially zero release.

Although this book deals with reactor safety and places emphasis on licensing only as a means of showing that safety is established, it is nevertheless appropriate here to comment outside these terms of reference. By detailing possible emissions from fossil-fuel plants, it is possible to view the very minor nuclear plant emissions in a more balanced light during the licensing process.

6.4.3.1 *Fossil Power Plant Releases*[†]

The principal fossil fuels, bituminous coal, natural gases, and oil, give off a variety of pollutants that include carbon monoxide, oxides of sulfur and nitrogen, hydrocarbons, and particulate matter. Of these, sulfur dioxide is the most critical.

[†] See Wright (*8*), Fish (*13*) and reference (*9a*).

Oil and gas also give rise to large quantities of nitrogen dioxide, while coal and oil contribute fly ash particulates. The exposure to these pollutants is measured in an individual by the action on his lungs, and in *some* cases, standards apply which limit the concentration of the pollutants in the air.

To place the emission of fossil-fuel pollutants in perspective, it is worth noting (*13*) that the discharge quantities of sulfur dioxide amount to between 27 lb/MWe-yr for gas and 306,000 lb/MWe-yr for coal, with oil coming in between. These figures, when compared to the concentration standard of 0.3 parts per million, imply that a dilution of the pollutant to the required standard may need up to nearly $2 \cdot 10^{11}$ m³ of air/MWe-yr (see Table 6.4).

In contrast, a PWR plant may discharge $5–10 \cdot 10^3$ μCi/MWe-yr. In order to dilute this to the concentration standard required by the AEC 10 CFR 20 regulations, 10^{-7} μCi/cm³, approximately $5–10 \cdot 10^4$ m³ of air/ MWe-yr is needed. In other words, the required dilution is something greater than a million times easier for the nuclear plants.

TABLE 6.4

DILUTION AIR REQUIRED TO MEET CONCENTRATION STANDARDS FOR VARIOUS POWER
PLANT POLLUTANTS[a]

Type of plant	Critical pollutant	Concentration standards	Discharge quantities per MWe-yr	Dilution air required per MWe-yr (m³)
Coal	SO₂	0.3 ppm	306,000 lb	$1.77 \cdot 10^{11}$
	Fly ash			
	²²⁶Ra	10^{-13} Ci/cm³	17.2 Ci	$1.72 \cdot 10^8$
Oil	SO₂	0.3 ppm	116,000 lb	$6.75 \cdot 10^{10}$
	Fly ash			
	²²⁶Ra	10^{-13} Ci/cm³	0.15 Ci	$1.5 \cdot 10^6$
	NO₂	2 ppm	47,000 lb	$5.77 \cdot 10^9$
Gas	SO₂	0.3 ppm	27 lb	$1.5 \cdot 10^7$
	NO₂	2 ppm	26,600 lb	$3.22 \cdot 10^9$
Nuclear	⁸⁵Kr	10^{-7} Ci/cm³	$5.7 \cdot 10^3$ Ci	$5.7 \cdot 10^4$ [b]
	¹³¹Xe		$9.5 \cdot 10^3$ Ci	$9.3 \cdot 10^4$ [c]

[a] See Fish (*13*).
[b] Shippingport 5 yr average.
[c] Yankee, 1965.

The RBE for radium and thorium from fossil-fuel plants is much higher than for the nuclear plants' krypton and iodine, therefore only a little radium release can have a relatively large effect. This is exhibited in Table 6.4 and also by the dose rates quoted for three power plants: Connecticut Yankee PWR (1968) gave $1.2 \cdot 10^{-6}$ μrem/hr-MWe, Dresden I BWR (1968) gave $8.7 \cdot 10^{-2}$ μrem/hr-MWe, while Widows Creek fossil plant gave $3.5 \cdot 10^{-5}$ μrem/hr-MWe. Thus the fossil plant gave just as much radioactive emission as nuclear plants even though it relied on an 800 ft stack to dilute effluent by a factor of 100 (*9b*).

Fast reactors do not differ very significantly from the LWR plants although they do discharge less and will have no difficulty in bettering the already excellent PWR standard.

The discharge of pollutants from fossil-fueled plants is being reduced. Effective methods for the control of particulates already exist and are in use. A Dolomite system for the removal of sulfur dioxide involves using finely divided limestone in a combustion chamber (*13*). Following a wet scrubber treatment, using an aqueous suspension of limestone or lime particles, it is hoped that up to 80% of the sulfur may be removed. This system is now being tested on some operating power plants. Significant reductions in the nitrogen oxides will require new combustion processes; in the meantime plants are being built with taller stacks to increase the dilution factor for the remaining discharged pollutants.

6.4.3.2 *Nuclear Plant Releases*[†]

A nuclear plant has to conform to strict release limitations set by the AEC in its Code of Federal Regulations outlined in the previous chapters. Moreover, this compliance is not measured against actual releases, as in the case of the fossil-fueled plants but against hypothetically bad conditions described as the design basis for the plant.

Thus Table 6.5 compares the annual doses which are recommended as maxima by the Federal Radiation Council against those that are calculated to occur if 1% of the fuel were leaking and also against those that are actually expected on the basis of present operating experience in PWR plants.

The table shows that the design basis conditions are about a hundredth of the recommended maxima or better at the various exclusion zone boundaries, while the actual expected values are again a further factor of a hundred lower (*8*).

[†] See Wright (*8*).

TABLE 6.5

ANNUAL DOSES FROM SINGLE NUCLEAR PLANT[a]

Dose (mrem/yr)	Medium[b]	Site boundary	Low population zone (5 miles)	General population zone (20 miles)
FRC recommended maximum	Air and water	500.0	170.0	170.0
Design basis	Air	5.0000	0.1040	0.0156
	Water	0.2055	0.0103	0.0093
	Total	5.2055	0.1143	0.0249
Actual expected	Air	0.0063	0.0001	0.0000
	Water	0.0435	0.0022	0.0020
	Total	0.0498	0.0023	0.0020

[a] See Wright (8).
[b] The water value includes the food chain contribution.

The assumptions in this table are that a 45-day hold-up system is used to eliminate all short-lived isotopes and that the gaseous releases are those resulting from pessimistic wind dispersal conditions. The liquid releases here are calculated from the intake of aquatic food as well as drinking water. The fish and mollusks account for about 5% of the total.

If multiple plants in one location are considered, then sample calculations showed that, for pessimistic conditions downstream, a set of three plants resulted in about a 50% increase in dose (8). Combined doses are not additive, unless the plants are in identical positions in relation to winds and ground water flows.

It is worth noting that the FRC limits apply only at the boundary of each zone, and the doses will be significantly less elsewhere farther away. Figure 6.7 shows that while the design basis release is much lower than the FRC exposure limit at the fence line, as one moves away the dose becomes exponentially less. In fact, even with a complete failure of the fuel in this example, where the FRC exposure limit would be exceeded at the fence line, doses drop below the limitation very rapidly as the distance is increased.

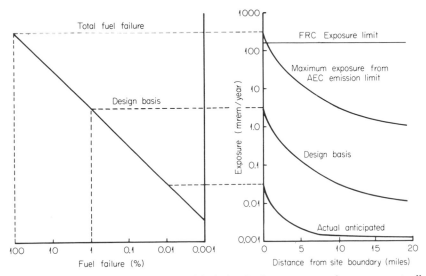

Fig. 6.7. Exposure limits compared with design basis exposures and exposures actually expected (*8*).

6.4.3.3 *Long Term Build-Up of Nuclear Pollutants*[†]

Concern has been expressed regarding the long term build-up of long-lived isotopes within the atmosphere even though individual discharges are well within the recommended limits. This concern arises from discharges from fuel processing plants rather than from the nuclear power systems, but nevertheless a word of clarification is in order.

During fuel reprocessing the isotopes are removed with the exception of xenon, tritium, and krypton. However, after a 45-day hold-up to allow short-half-life isotopic decay, the xenons are removed and only tritium and ^{85}Kr are released. They have half-lives of 10 and 12 yr and they therefore collect uniformly throughout the biosphere. What levels might these isotopes reach if nuclear expansion continues at its expected rate?

Figure 6.8 shows that the planetary build-up of tritium, assuming projected nuclear power expansion, attains about 100 million Ci by the year 2000. Although this level sounds high, it is in fact about 6% of the maximum tritium activity due to weapons fallout at its peak in 1963 and it is only just equal to the natural production of tritium from the sun's cosmic rays. The planetary exposure of the world's population at the year 2000 would be about 0.002 mrem/yr. This assumes that half the world's power by this year would be produced from nuclear plants.

[†] See Wright (*8*) and reference (*9b*).

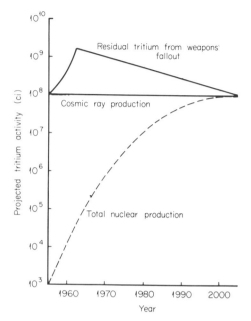

Fig. 6.8. Cosmic ray production of tritium in the earth's atmosphere provides an equilibrium value of 100 million Ci. Estimated tritium activity produced by nuclear power facilities in the year 2000 will also be 100 million Cu, or approximately 6% of the maximum tritium activity present in 1968 (*8*).

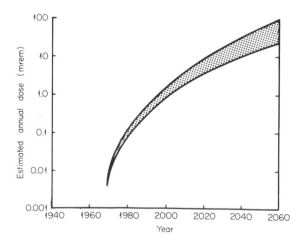

Fig. 6.9. Within a century, average exposures to the world population from [85]Kr could reach 50–100 mrems if retention systems are not developed for fuel processing plants (*8*).

The case of ^{85}Kr is a slightly different matter because the atmospheric content is produced entirely from nuclear reactions. The present dose is less than 0.1% of the combined background due to cosmic rays and other natural background. However, Fig. 6.9 shows that by the year 2060 the accumulated krypton will increase to about a level which would give an exposure of approximately 1000 times that of tritium, with an annual individual dose of 50–100 mrem/yr. While this dose is still not dangerous, the accumulation does imply that some retention schemes should be employed in fuel processing plants to avoid this accumulation. Present technological capabilities promise effective future means of controlling krypton discharges by cryogenic concentration (*10*).

This short review of discharges into the air from fossil-fueled and nuclear-fueled plants shows that the nuclear plants are subjected to far greater control in discharge rates, and indeed they have less of an immediate problem. Moreover, in the long run, biospheric accumulation from fuel processing plants, although the subject of sensation-seeking journalism, is not a problem. Even so, improvements in the management of ^{85}Kr are to be expected.

REFERENCES

1. Nuclear vessels. "ASME Boiler and Pressure Vessel Code," Sect. 3. ASME, New York, 1968.

2a. Proposed IEEE criteria for nuclear power plant protection systems. IEEE-279, August 30, 1968.

2b. Proposed IEEE criteria for class IE electrical systems for nuclear power generating stations. Rept. No. NSG/TCS/SC4-1, IEEE, June 1969.

2c. Proposed Joint Committee on Nuclear Protection Standards guides, 1970.

3. Code of Federal Regulations—Sect. 10, Atomic Energy Law Reports. Pt. 20; Standards for Protection against Radiation (January 1961), Pt. 50: Licensing of Production and Utilization Facilities (February 1969), and Pt. 100: Reactor Site Criteria (April 1962).

4. 1966 Book of Standards in 32 parts. *Book ASTM Stand.* (1966).

5. W. B. Cottrell, "Compilation of United States Nuclear Standards." ORNL-NSIC-57. 5th ed. U. S. Amer. Stand. Inst., Oak Ridge Nat. Lab., Oak Ridge, Tennessee, 1968.

6. General design criteria for nuclear power plants. 10 CFR 50, Code of Federal Regulations, Fed. Regist., February 10, 1971.

7a. "Licensing of Power Reactors." AEC Booklet, April 1967.

7b. "A Guide for the Organization and Contents of Safety Analysis Reports." USAEC, TID-24631 (REG. 1), June 30, 1966.

7c. Winfrith SGHWR opening ceremony. *Atom* **138** (1968).

8. J. H. Wright, Electrical power generation and the environment. *Westinghouse Eng.* **30** (3), 66 (1970).

9a. "Nuclear Power and the Environment" (Understanding the Atom ser. of booklets). U. S. At. Energy Comm. Div. of Tech. Inform., Oak Ridge Nat. Lab., Oak Ridge, Tennessee, 1969.

9b. "Environmental Effects of Producing Electric Power." Hearings before Joint Comm. At. Energy, October and November 1969, Part I, U. S. Government Printing Office, Washington, D.C., 1969.

10. E. Thro, The controversy over thermal effects. *Nucl. News*, p. 49, December 1968.

11. "Problems in the Disposal of Waste Heat from Steam-Electric Plants." Bur. of Power, Federal Power Comm., 1969.

12. AEC. Statement on Environmental Policy Act of 1969, Atomic Energy Clearing House, **16** (14), pp. 2–5, April 1970.

13. B. R. Fish, Radiation in perspective—The role of nuclear energy in the control of air pollution. *Nucl. Safety* **10**, (2), (1969).

14. E. W. Fowler and D. E. Voit, "A Review of the Radiological and Environmental Aspects of Krypton-85." U. S. Dept. of Health, Education, and Welfare, Public Health Service, Washington D.C., September 1969.

APPENDIX

COMPUTER CODES FOR NUCLEAR ACCIDENT ANALYSIS

The following set of codes represents a partial list of digital computer codes that are used in the analysis of aspects of accidents in liquid-metal-cooled fast reactors. Many of the codes are available through the Argonne Code Center.[†]

Some of the codes listed are proprietary to their corporate authors, and others exist that are not listed because they are not available, not yet complete, or not fully documented.

[†] Address: Argonne National Laboratory, 9700 South Cass Avenue, Argonne, Illinois 60439.

Name of code	Function	Language[a]	Reference[b]
FORE II	Calculates neutron kinetics and thermal behavior of up to 3 reactor channels at 7 axial positions	FIV	*1*
IANUS	Calculates kinetics, thermal behavior of the core, and hydraulic behavior of the primary and secondary loops of a loop-type LMFBR	FIV	*2*
TOSS	Solves heat conduction for a 3-dimensional mesh which may include a fuel rod under nonsymmetric cooling at the boundary (see TRUMP)	FIV	*3*
PRANC	Calculates hydraulics in a multiloop primary coolant circuit following a rupture, and includes pump cavitation	FIV	*4*
COBRA, HAMBO	These codes calculate heat transfer in a fuel assembly allowing for coolant cross flow between subchannels depending on the lateral resistance and the dimensions. Both codes were originally designed to include water–steam mixtures	FIV	*5, 6, 7*
SAMBA	An extension of HAMBO to allow for a subchannel blockage	—	—
PASET	Calculates sodium voiding within a subchannel using an annular model for the voiding. The effects of nuclear feedback and primary coolant loop hydraulic feedback are included	FIV	*8*
SAS-1A	Comprehensive dynamics code including a detailed fuel and cladding model, sodium boiling in two-phase and annular regimes. Multi-channel	F	*9*
FIGAFRO	Computes fission gas outflow from a ruptured fuel rod as a function of optional gas flow paths from the fission product plenum through sintered and unsintered fuel regions to the cladding rupture. Includes both sonic and subsonic gas flow.	FIV	*10*

Name of code	Function	Language[a]	Reference[b]
FIDES	Computes molten fuel ejection from a ruptured fuel pin subject to the inertial mass of molten fuel, the pressure in the fuel pin, and that in the coolant channel (PASET includes a simple FIDES model and includes allowance for fuel–sodium interaction)	FIV	*11*
MELT II	Computes the slumping of fuel and the reactivity feedback due to that slumping, dependent on the slumping model chosen by the user	—	*12*
AX-I	Calculates energy yield for fast reactor excursion	F63	*13*
AX-TNT	Solves hydrodynamic, thermodynamic, and neutron kinetic coupled equations to obtain yield of a fast reactor dispersion. Includes a calculation of the damaging available work	F63	*14*
COOT, EXTRA, PANIC	Various energy yield code calculations	—	*15, 16, 17*
MARS	Two-zone perturbation calculation of the energy release following a reactivity insertion. Has been variously modified to include a calculation of the work energy available depending on the quantity of sodium within the core	FIV	*18*
VENUS	Two-zone hydrodynamic calculation of the energy release following a reactivity insertion. It can calculate the effects of an implosion with large positional changes of fuel	F	*19*
SHAMAN	Calculates vessel radial strain from chemical explosion correlation due to Proctor and allows for sodium hammer impact with head	FIV	*20*
HEADAKE	Computes the sodium hammer which results after successive waves of hot fuel are driven into the sodium above the core, following a disruptive accident	FIV	—
HEDLIFT	Computes bolt stretch and head lift as a result of the sodium hammer produced by the HEADAKE code	FIV	—

Name of code	Function	Language[a]	Reference[b]
REXCO	Solves a Lagrangian set of equations describing the hydrodynamics of the shock and pressure waves following a core disruptive accident, from an initial energy and pressure distribution. No heat transfer is included yet. REXCO-H is the basic program, REXCO-I includes the effect of inelastic response of the vessel walls	FIV	*21, 22*
FEATS	Two-dimensional finite element program for computing thermal stress and mechanical stress due to a variety of boundary conditions (one of a large number of similar programs)	FIV	*23*
CONTENT	Predicts the temperature–pressure response of a containment volume due to the deposition of heat as a distributed aerosol or a deposited mass of debris	FIV	*24*
AISITE II	Determines siting criteria for reactors. Computes dose versus distance, depending on engineering parameters and meteorological conditions and relates these to doses to real organs	FII	*25*
PREP & KITT	Codes for the automatic evaluation of fault trees by Boolean algebra. PREP evaluates the minimal path sets (success modes) and minimal cut sets (failure modes) in preparation for KITT which evaluates the numerical probabilities	—	*26*
TRUMP	Solves the heat conduction for a three-dimensional mesh which may include a fuel rod under nonsymmetric cooling at the boundary. Does the same job as TOSS but allows a variable coolant temperature boundary	—	*27*
SOFIRE II	A containment code to assess the pressure and temperature effects of a sodium fire in a building	—	*28*

[a] The letter F refers to one of the variety of the FORTRAN languages that is then specified by number.

[b] The corporate author is made clear by the origin of the reference.

REFERENCES

1. J. N. Fox, B. E. Lawler, and H. R. Butz, FORE II: A computational program for the analysis of steady state and transient reactor performance. USAEC Rept. GEAP-5273. General Elect. Co., San Jose, California, September 1966.

2. R. Schmidt, F. M. Heck, and C. F. Wolfe, The significance of heat capacities and time delays on the thermal transient requirements for LMFBR components. *Trans. Amer. Nucl. Soc.* **11** (2), (1968).

3. B. L. Pierce, Modified transient and/or steady state (TOSS) digital heat transfer code. WANL-TMI-1020, Westinghouse Astronuclear Lab. Pittsburgh, Pennsylvania. April 1964.

4. E. F. Beckett, J. A. George, R. W. Tilbrook, and J. A. Zoubeck, Response of a piped LMFBR to primary system pipe rupture. *1st. Annu. Conf. ASME Nucl. Power Div., Palo Alto, California,* ASME Paper, 1971.

5. D. S. Rowe, Cross-flow mixing between parallel flow channels during boiling. Pt. I, COBRA: Computer program for coolant boiling in rod arrays. BNWL-371 (Pt. 1), March 1967.

6. D. S. Rowe, COBRA-II: A digital computer program for thermal-hydraulic sub-channel analysis of rod bundle nuclear fuel elements. BNWL-1229, February 1970.

7. R. W. Bowring, HAMBO: A computer programme for the sub-channel analysis of the hydraulic and burnout characteristics of rod-clusters, Pt. 1, General description. Brit. Rept. AEEW-R/524, A.E.E. Winfrith, Dorset, England. April 1967.

8. R. W. Tilbrook and G. Macrae, PASET: A transient code for plant analysis of sodium voiding. *Trans. Amer. Nucl. Soc.* **12**, 904 (1969).

9. D. R. MacFarlane, ed., SAS-IA: A computer code for the analysis of fast reactor power and flow transients. ANL-7607, Argonne Nat. Lab., Argonne, Illinois. 1971.

10. M. D. Carelli and R. D. Coffield, Fission gas released from failed fuel pins in liquid metal fast breeder reactors. WARD 5446, Westinghouse Advanced Reactors Division, Waltz Mill, Pennsylvania. September 1970.

11. M. D. Carelli, Fission gas and molten fuel ejection from failed fuel rods in LMFBRs. *Amer. Nucl. Soc. Topical Meeting New Develop. Reactor Math. Appl. Idaho Falls, Idaho, March 1971.*

12. A. E. Waltar, A. Padilla, and R. J. Shields, MELT-II: A two-dimensional neutronics–heat transfer computer program for fast reactor safety analysis. WHAN-FR-3. WADCO, Richland, Washington. 1970.

13. D. Okrent *et al.,* AX-I, a computing program for coupled neutronics hydrodynamics calculations on the IBM-704. USAEC Rept. ANL-5977. Argonne Nat. Lab., Argonne, Illinois, 1959.

14. C. J. Anderson, AX-TNT, A code for the investigation of reactor excursions and blast waves from a spherical charge. USAEC Rept. TIM-951. Pratt and Whitney Aircraft, Middletown, Connecticutt. September 1965.

15. L. M. Russell and E. Morris, A computer calculation of the energy yield produced by a hypothetical fast reactor accident. *Proc. Conf. Appl. Comput. Methods to Reactor Problems, A.N.L., May 17–19, 1965,* USAEC Rept. ANL-7050, pp. 503–514. Argonne Nat. Lab., Argonne, Illinois, August 1965.

16. R. M. Lord, Effect of core configuration on the explosive yields from large sodium-cooled fast reactors. *Proc. Int. Conf. Safety of Fast Reactors, September 19–22, 1967, Aix-en-Provence, 1967.* Paper III-3-1.

17. A. Renard and M. Stievenart, Evaluation of the energy release in case of a severe accident for a fast reactor with important feedbacks. *Proc. Int. Conf. Safety of Fast Reactors, September 19–22, 1967, Aix-en-Provence, 1967.* Commissariat à l'Énergie Atomique, Paris. 1967.
18. N. Hirakawa, MARS: A two dimensional excursion code. APDA-198, June 1967.
19. W. T. Sha and T. M. Hughes, VENUS: Two dimensional coupled neutronics-hydrodynamics coupled computer program for fast reactor power excursions. ANL-7701, 1970.
20. Method of J. F. Proctor, Adequacy of explosion-response data in estimating reactor-vessel damage. *Nucl. Safety* **8** (6), 565 (1967).
21. Y. W. Chang, J. Gvildys, and S. H. Fistedis, Two dimensional hydrodynamics analysis for the primary containment. ANL 7498. Argonne Nat. Lab., Argonne, Illinois, June 1969.
22. G. Cinelli, J. Gvildys, and S. H. Fistedis, Inelastic response of primary reactor containment to high-energy excursions. ANL-7499. Argonne Nat. Lab., Argonne, Illinois, August 1969.
23. J. A. Swanson, FEATS: A computer program for the finite element thermal stress analysis of plane or axisymmetric solids. WANL-TME-188, Westinghouse Astronuclear Lab., Pittsburgh, Pennsylvania. January 1969.
24. J. A. Zoubeck, Private communication, Westinghouse Advanced Reactors Division, Waltz Mill, Pennsylvania. 1970.
25. R. A. Blaine and E. L. Bramblett, AISITE II: A digital computer program for investigation of reactor siting. USAEC Rept. NAA-SR-9982. Atomics Int., Canoga Park, California. October 1964.
26. W. E. Vesely and R. E. Narum, PREP and KITT: Computer codes for the automatic evaluation of a fault tree. USAEC Rept. IN-1349, Idaho Nuclear Corp., Idaho Falls, Idaho, August 1970.
27. A. L. Edwards, TRUMP: A computer program for transient and steady state temperature distributions in multidimensional systems. UCRL-14754, Revision 1, University of California Radiation Laboratory, Berkley, California. May 1968.
28. F. P. Beiriger, SOFIRE II: Engineering code users manual. Atomics Int. Internal Document, TI-707-14-011. Atomics Int., Canoga Park, California. 1970.

INDEX